T0155992

SpringerBriefs in Climate Studies

More information about this series at http://www.springer.com/series/11581

Antonio A. Romano • Giuseppe Scandurra
Alfonso Carfora • Monica Ronghi

Climate Finance as an Instrument to Promote the Green Growth in Developing Countries

 Springer

Antonio A. Romano
Department of Management Studies
and Quantitative Methods
University of Naples "Parthenope"
Naples, Italy

Giuseppe Scandurra
Department of Management Studies
and Quantitative Methods
University of Naples "Parthenope"
Naples, Italy

Alfonso Carfora
Italian Revenue Agency
Rome, Italy

Monica Ronghi
Department of Management Studies
and Quantitative Methods
University of Naples "Parthenope"
Naples, Italy

ISSN 2213-784X   ISSN 2213-7858   (electronic)
SpringerBriefs in Climate Studies
ISBN 978-3-319-60710-8   ISBN 978-3-319-60711-5   (eBook)
DOI 10.1007/978-3-319-60711-5

Library of Congress Control Number: 2017949179

Printed on acid-free paper

This Springer imprint is published by Springer Nature
The registered company is Springer International Publishing AG
The registered company address is: Gewerbestrasse 11, 6330 Cham, Switzerland

Contents

Abbreviations

UNFCCC	United Nations Framework Convention on Climate Change
GHG	Greenhouse gas emissions
COP	Conference of the Parties
USD	US Dollar
SIDS	Small Island Developing States
LDCs	Least developed countries
GCF	Green Climate Fund
OECD	Organization for Economic Cooperation and Development
EIT	Economies in transition
GEF	Global Environment Facility
CIFs	Climate Investment Funds
AF	Adaptation Fund
UNDP	United Nations Development Programme
UNEP	United Nations Environment Programme
NGOs	Non-governmental organizations
RES	Renewable energy sources
ODA	Official development assistance
DAC	Development Assistance Committee
OOF	Other official flows
CP3	Climate Public Private Partnership
GNI	Gross national income
EPI	Environmental Pollution Index
JICA	Japan International Cooperation Agency
JPP	Japan Partnership Program
CAIT	Climate Analysis Indicators Tool
GDP	Gross domestic product
CO_2	Carbon dioxide
IEA	International Energy Agency
CH_4	Methane

N_2O	Nitrous oxide
CFCs	Chlorofluorocarbons
HCFCs	Hydrochlorofluorocarbons
HFCs	Hydrofluorocarbons
PFCs	Perfluorocarbons
SF6	Sulfur hexafluoride
CIs	Composite indicators
BOD	Benefit of the doubt
MPI	Mazziotta-Pareto index
EW	Equal weighting
PCA	Principal component analysis
PCs	Principal components
LAD	Least absolute deviation
EIA	US Energy Information Administration

List of Figures

List of Tables

Chapter 1
Introduction

Abstract In this chapter we introduce the topics and the goals of the book. We present the problem of environmental sustainability of the current economic production system and consumption of goods starting from one of the first approach attempts through dynamical systems proposed by J.W. Forrester. It shows how concern for the Earth system was already important in the late '60s. Finally, it outlines the work plan in view of the deepening of a measure of adaptation-mitigation as the Climate Finance.

Keywords System dynamics models • Stock and flows systems • Sustainability • Conferences of parties

1.1 Introduction and Background

Since the beginning of 1972, the expected development of some of the most important social, demographic, environmental and economic variables that form the foundation of growth and human development is been outlined in the book on *"The Limits to Growth"* by Donella H. Meadows et al. (1972). A summary, necessarily incomplete, of the results obtained is shown in Fig. 1.1, in the uncertain graphic of the calculators in those years. For those variables, the work included a number of points of no return, distributed between 2015 and 2050.

It is unnecessary today to assess the accuracy of the forecasts[1] made by the authors at the time, but it is interesting to point out at least two important points: (*i*) the fact that even in those years, the problem of the sustainability of economic development, left to itself, had become so important as to draw the attention of organizations such as the Massachusetts Institute of Technology (MIT), the Club of Rome and the Volkswagen Foundation; and (*ii*) for the proposed method in order to obtain predictions, that approached for the first time, the centre stage, in a field that

---

[1]Dynamic models do not make forecasts in the common sense of the term. They once built and tested the model, show future trends of variables involved in the hypothesis that the structure of the model remains unchanged over time. The model can also consider the possibility of including exogenous factors but its structure should not change.

© The Author(s) 2018
A.A. Romano et al., *Climate Finance as an Instrument to Promote the Green Growth in Developing Countries*, SpringerBriefs in Climate Studies, DOI 10.1007/978-3-319-60711-5_1

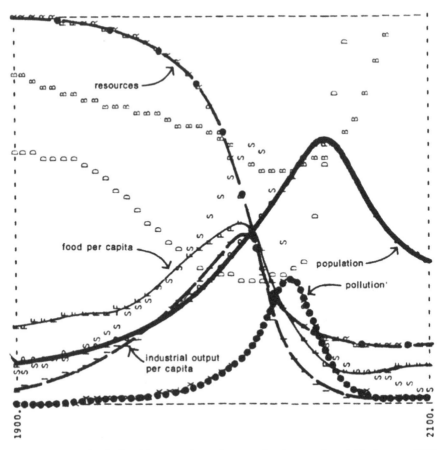

Fig. 1.1 Expected path of social, economic, environmental and economic variables between 1990 and 2100. (Source: D.H. Meadows et al. 1972)

was different to that of Industrial Controls: it was one of the first applications of System Dynamics developed by J.W. Forrester of the MIT. This scientific approach is the indispensable fulcrum for the preparation of complex scenarios that can be constructed by acting independently on each singular variable of the model, while leaving unchanged the rest of the system. This is a key factor, both with regards to being able to measure the effect of global warming in laboratory tests that are not possible to perform in the real world, as well as in relation to providing a possible solution to the problem of the attribution of the effects to a natural or anthropogenic cause.

The System Dynamics models, used synergistically with other scientific approaches and with the use of broad-based knowledge and observations stemming from very different sources, are today still used extensively for data analysis and for their ability to offer some reasonable predictions assuming, naturally, that the structure of the Earth's climate system, maintains, in time, its internal coherence represented and tested in different models.

It should be noted that the study presented in 1972, did not directly concern the phenomenon of global warming, nor is it possible that the technological and informational means adopted at the time be in the least comparable to those of today. However, it is not remarkable that the findings underpinning the study, which the authors come to in a completely different way, lean towards the unsustainability of the development model, still ongoing today, at least under the assumption of *business-as-usual*. From this point of view, the methodological tool proposed in the late 60^s still keeps intact today both the basic approach (a system of stocks and flows) as well as its practical effectiveness (attribution of effects and forecasts bound to the stability of the structure of the system).

That the world has the obligation to find solutions to mitigate global warming and prepare the necessary measures to adapt to the changes, already partly irreversible, is indisputable even for those who deny the role played by mankind in causing this phenomenon. U Thant, UN Secretary General in the 47th Session of the United Nations in 1969 (UN 1969) wrote "... *for the first-time in the history of mankind there is arising a crisis of world-wide proportions involving developed and developing countries alike, - the crisis of the human environment. Portents of this crisis have long been apparent – in the explosive growth of human population, in the poor integration of a powerful and efficient technology with environmental requirements, in the deterioration of agricoltural lands, in the unplanned extension of urban areas, in the decrease of available space and the growing danger of extinction of many forms of animal and plant life. It is becoming apparent that if current trends continue, the future of life on earth could be endangered*". The same secretary, for greater clarity, added: "*I do not wish to seem overdramatic but I can only conclude from the information that is available to me as Secretary-General that the Members of the United Nations have perhaps ten years left in which to subordinate their ancient quarrels and launch a global partnership to curb the arms race, to improve the human environment, to defuse the population explosion, and to supply the required momentum to development efforts. If such a global partnership is not forged within the next decade, then I very much fear that the problems I have mentioned will have reached such staggering proportions that they will be beyond our capacity to control.*" (Thant 1969).

In fact, it was necessary to wait until 11th December 1997 to see the signing of a first real, albeit partial, Memorandum of Understanding between the parties (COP3), aimed at combatting global warming. The Protocol entered into force only in the year 2005 and was followed by other agreements up until the meeting in Paris, at first, and then subsequently, the Marrakesh Declaration, approved in 2016 by 196 countries that participated in the UN Climate Change Conference, COP22.

The agreements reached are considered by many scholars to be still insufficient, even for achieving the goal of keeping global warming within the limit of 2 °C, compared with the pre-industrial era, by the end of the century. To achieve this objective, there will necessarily be the need to proceed with measures to reduce greenhouse gas emissions, improve the efficiency of production processes, reduce the energy intensity of production, improve the production of energy and the need to find appropriate ways to ensure the development of the poorest countries via production systems that are compatible with environmental objectives.

On the other hand, given the long lifetime of greenhouse gases, we must put into place interventions that can absorb the direct and indirect impact caused by global warming with appropriate targeted adaptation measures, especially for the most vulnerable countries and those who reside along the coastal areas.

In particular, this work is dedicated to a particular series of measures which, at the same time, both belong to the field of mitigation measures as well as to those in terms of adaptation. In fact, it is well-known that since 1992 the United Nations has highlighted the need to accumulate new financial resources to support the development of poorer countries. These resources, constituted in the richest countries, form the cash flow that finances economic activities in the developing countries. These activities, in turn, in addition to promoting development, must promote a planning strategy that is capable of meeting the goal of mitigation and adaptation of / to global warming and its effects.

Much has been written with regards to how to provide the resources in question and the fields of activity that these resources finance. This work is one of the few attempts to evaluate, ex post, the distribution of these financial flows and the effectiveness they have had in facilitating the achievement of the objectives.

The second chapter presents a quick overview on the problem of climate change, its evolution over time and the question of the attribution of the impacts of global warming. This is followed by a presentation of the main mitigation and adaptation measures derived from reports from the most accredited bodies.

The third chapter, focuses, in particular, on *Climate Finance* as it appears from the meetings that have taken place from the COP1 to the COP22, the background picture in which the interventions to date have evolved and, above all, a first analysis of the financial flows, both with regards to their entities, as well as their origin and their destination, will be made.

The fourth chapter, is dedicated to both the proposal of a metric suitable for modelling, both with regards to the construction of a possible synthetic indicator capable of reducing, to a dimension, a phenomenon that, by its nature, manifests peculiar multidimensional characteristics. Moreover, the quantile regression is presented, focusing on the theory and possible applications in research fields of interest. Quantile regression, in fact, is one of the tools used to assess the effects of economic policies on some of the target variables.

The fifth chapter proposes one of the possible approaches on the effectiveness of the application of Climate Finance measures. After a description of the observed data and assumptions used for the application of the methodology, there will follow a construction of a composite indicator capable of providing a quantitative measure of the environmental performance of a country related to other variables that are able to describe social and economic particularities. In this way, it is possible to individuate a way capable of giving a measure of the repercussion of the financial flows on combating environmental degradation seen through the most important components of the increase in global warming.

Finally, in the sixth chapter, we will discuss the results of the study and draw some conclusions for a future, more detailed analysis.

The Appendices outlines some methodological notes, tools and data sources that can be used by readers who wish to retrace the steps followed by the authors.

References

Meadows DH, Meadows DL, Randers J, Beherens WW (1972) The limits to growt, a report for the club of Rome. Universe Books, New York

Thant U (1969) quotation cited in R. Theobald, Challenge of a decade: global development or global breakdown, pamphlet prepared for the United Nations Centre for Economic and Social Information, New York

United Nation (1969) Problems of the human environment, 47° session, E/4667, New York

Chapter 2
Climate Change

Abstract This chapter deepens the thermal balance of the Earth-Atmosphere System with respect to the balance between the incoming solar energy in the system and that the outgoing radiation heat. It introduces the concept of the greenhouse effect – analyzing the natural and the enhanced greenhouse effect – and outlines the role that the gases contained in the atmosphere have in determining the global average temperature of the Earth's surface. It also introduces the concept of reaction and the important influence that it has in the evolution of thermal equilibrium of the Earth System. From the scientific evidence of global warming, also outline the requirements to mitigate the consequences of climate change and, given the long permanence of greenhouse gases in the atmosphere, to provide measures for adaptation to the effects of global heating.

Keywords Radiation • Solar energy • Global mean energy balance • Climate change • Global warming • Greenhouse gas • Natural greenhouse effects • Enhanced greenhouse effect • Adaptation • Mitigation

2.1 Evolution and Path

The *Earth system*, within which all the phenomena that will be discussed later take place, is composed of five domains: the lithosphere, the hydrosphere, the cryosphere, the atmosphere and the biosphere. The lithosphere is the solid part of the planet located in the initial kilometres of depth of the entire geosphere; the hydrosphere is the part of the globe that contains water in all its states, liquid, solid and gaseous; the atmosphere is the gaseous envelope of the planet inside which the gases are distributed in chemical and physical forms and in different concentrations; the biosphere is the domain where life develops and unfolds, including human life. In the first four domains, the balance between incoming energy (literally solely solar energy) and that emerging from the system, is determined and takes shape. Thanks to the structure of the Earth System, the planet maintains (or should maintain) chemical, physical and thermal conditions suitable for life as we know it and as we understand it.

© The Author(s) 2018
A.A. Romano et al., *Climate Finance as an Instrument to Promote the Green Growth in Developing Countries*, SpringerBriefs in Climate Studies, DOI 10.1007/978-3-319-60711-5_2

These conditions have not always been stable in the millions of years of life on Earth but have varied, even to a great extent, due to natural phenomena, primarily if not exclusively, linked to solar activity, the inclination of the Earth axis and the orbiting of the planet around the Sun. From this point of view, the planet Earth is "accustomed" to climate change constituted by cyclical transitions from glacial climates to warmer periods (inter-glacial), such as the one we are now experiencing (the Holocene). The fact is that humankind began to affect the climate since the advent of farming and population growth. In particular, since the beginning of the so-called "Industrial Age" (dating from the mid-eighteenth century), the human footprint has become significantly stronger, to the point that the vast majority of scientists believe that the initiation of the planet's passage into a new Ice Age was interrupted and reversed by the impact of human activities (including the huge population growth) on the thermal balance of the System. Naturally, the effect is that of a distinct global warming of the entire planet. This is why we often speak about Global Warming as a synonym for Climate Change, but in reality the two concepts do not indicate the same phenomenon (see, eg, Hansen et al. 2010).

The United Nations Framework Convention on Climate Change (UNFCCC), therefore, uses the term climate change only to refer to changes produced by humankind and that of climate variability for changes resulting from natural causes. Due to the fact that climate change in the sense as indicated by the UNFCC, is manifested by an overall increase in the average surface temperature of the earth, thus *global warming* refers to *an unusual, rapid growth of the average surface temperature in the last century and in this century due mainly to the greenhouse gases issued by* mankind *as a result of the burning of fossil fuels*. Below, we will talk first of all about the natural greenhouse effect and the feedback concept in order to understand the mechanisms that permit the existence of life on the planet; then, we will go on to analyse the enhanced greenhouse effect to see how anthropogenic emissions cause global warming.

2.2 Natural Greenhouse Effect

Climate change and global warming are two expressions that are often used as an alternative to one another. This choice is not entirely justified but, in a first approximation, it can be considered acceptable.

The current chemical and physical conditions of the Earth System, making life possible as we know it, are largely due to the interactions between the solar energy affecting the planet (shortwave radiation ~ 0.3–1 micron), the atmosphere (typically the lowest part of it: the troposphere and the stratosphere), the hydrosphere (with its ability to store and distribute heat), and the continental surface (with its ability to emit long-wave radiation ~ 7–25 micron). In the continental part, an important role is played by the cryosphere which helps to reflect a non-secondary part of incident solar radiation and that influences the sea levels. Without the presence of the atmosphere, the average global temperature of the planet would stabilize at around

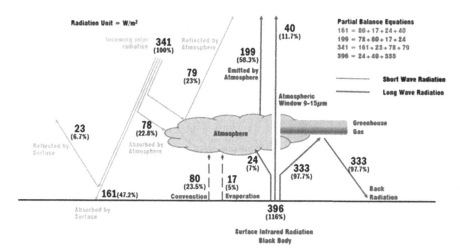

Fig. 2.1 The equations in the *upper* right show the items in the radiation balance of the Earth System. In *yellow* displays the components of the radiant energy from the sun. In *red* shows long-wave radiation emitted by the Earth due to the temperature reached by it. The 40 W/m^2 "pierce" the atmosphere and pass directly in space thanks to a "window" transparent to that frequency band. The natural warming effect of the planet issues from 333 W/m^2 that GHG reflect on the surface that emitted them

−15 °C with large daily and seasonal variabilities. A climate of this type would certainly not be suitable for life on Earth.

In fact, even our planet, just like any other black body with a temperature above 0° K, emits, to the atmosphere and then into space, infrared radiation (heat). For the wavelength of this radiation, the atmosphere is not transparent and it is mostly reflected again onto the Earth (Fig. 2.1). This reflected part is responsible for the natural greenhouse effect and causes the global mean surface temperature to stabilize around +15 °C with a variability that is quite acceptable for the life of plant and animal species on the Earth.

The atmosphere's ability to reflect long-wave radiation is due to the presence in the atmosphere of certain gaseous elements among which the most important are carbon dioxide (CO_2), methane (CH_4), water vapour (H_2O), nitrous oxide (N_2O) and sulphur hexafluoride (SF_6). The atoms within the molecules of these gases, when they are bombarded with radiation of an appropriate wavelength, begin to vibrate elastically and absorb energy that subsequently is re-emitted in every direction. Thus, while a part of infrared thermal energy is reflected directly back into space, another part, absorbed by the greenhouse gases, is reflected towards the surface. The balance between the part reflected towards space and the return part reflected towards the surface is determined by rules related to the mixture of the thermal wavelengths emitted from the Earth as a black body. Figure 2.2 illustrates how the gaseous elements in the atmosphere filter the different frequencies of heat emissions. It should be noted that the solar energy that enters the system through the upper limit of the atmosphere comes entirely from radiation in the form of

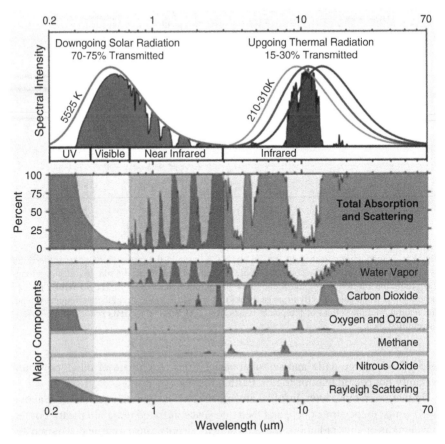

Fig. 2.2 Radiation emitted and absorbed by the Earth's surface and from the atmosphere. The first strip shows the frequency spectra (wavelengths) of radiation emitted by the Sun and the Earth. The second strip reports the total absorption spectrum and the spread of radiation operated by the various components of the atmosphere. The third strip shows the absorption-diffusion spectra attributable to each of the above components. Note the primary role of water vapor and that of carbon dioxide. The latter closes the "window" relative to the frequencies of greater intensity of the   emission   spectrum   of   the   Earth   (commons.wikimedia.org/wiki/File: Atmospheric_Transmission.png)

electromagnetic radiation which has frequencies in the range of ultraviolet, the visible and the near infrared. The initial part of the spectrum (ultraviolet) is partly absorbed by the ozone in the atmosphere and in part by the Rayleigh scattering and therefore does not reach the earth's surface. The infrared part of the spectrum is masked in areas by water vapour. The Heat radiation emitted by the Earth is instead entirely in infrared. For fields ranging between 0.5 and 1 micron and 8–10 micron, the atmosphere is transparent and the radiation of those frequencies is dispersed directly into space (the "window" of the Atmosphere). All the other radiation is absorbed and then reflected toward the surface, determining the afore-mentioned greenhouse effect[1] (Trenberth et al. 2009).

Within this mechanism, another important role is played by aerosols. These compounds are essentially formed from organic carbon, transition metals, ions etc. Their effect is to absorb and reflect sunlight and for this reason, in contrast to the properly so-called greenhouse gases, they tend to cool the atmosphere. The clear effect of all these elements is to generate that additional heating which raises the global average surface temperature to a value of about 15 °C, that is almost 30 °C higher than what the temperature would actually be without the atmosphere and its gases. What has been described is the energy exchange process (dynamic) that occurs when the balance between the incoming energy into the system and that which exits, is reached.

In fact, the Earth system never reaches a stable equilibrium, even in the absence of human activities. This is due to the superimposition of the effects of various thermal cycles of periods and variable amplitudes in time and in space. During these cycles, the level of solar radiation varies considerably and this results in a continuous imbalance of the above-described heat exchange mechanism. The Earth responds to these imbalances by varying its surface temperature in its continual search for a new and stable equilibrium. Each cause, internal or external to the Earth system, capable of disrupting the thermal balance, goes under the name of Radiative Forcing. Hence, also the variations in intensity of incident solar radiation must be considered radiative forcing, even if they are of a natural origin.

The cycles of solar activity have been grouped based on their average length: the best known are those daily and seasonal (quarterly) but to these must be added 10-year, century and pluri-century cycles. For example, in Fig. 2.3 some cycles characterized by different time scales, are reported: millennial cycles, whose performance has been reconstructed on the basis of paleoclimatology and 10-year cycles. The last graphic refers to natural non-cyclical variations due to volcanic eruptions. All of these variations, that overlap, cause clear effects that influence not only the surface temperature of the Earth directly, but also the concentration of greenhouse gases in the atmosphere due to the perturbations induced in the carbon cycle, water cycle, etc. In turn, these perturbations cause changes in the surface temperature. The concept of the chain influences (feedback), insufficiently treated by regressive techniques, deserves some consideration that will be given in the next section.

---

[1]The choice of the term "greenhouse effect" is unhappy. In fact, in greenhouses, the internal temperature increase is essentially due to the fact that the glazed surfaces eliminate the convective air effect by preventing the heat exchange with the outside. At equilibrium, the interior of the greenhouse and the glass surfaces will reach a new temperature limited by the convective motion of air around the same greenhouse which will eliminate the heat. In the case of the Earth System, instead, the heating mechanism is due to the difference between the energy of the incoming wavelength and the outgoing one.

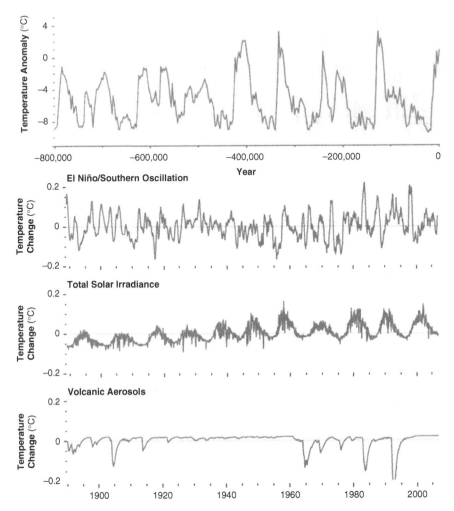

Fig. 2.3 Some examples of solar irradiation cycles and energy variation due to natural phenomena. Note the scale factor between the first graph and the other three. In the first of the three graphs it is cyclical trends that develop hundreds of thousands of years. The other three describe decennial phenomena. The first graph shows temperature anomalies due to small variations in Earth's orbit. These latter determine the sequence of ice ages (of longer duration) and interglacial eras (of shorter duration). The second graph shows temperature anomalies due to the periodic warming of Pacific Ocean waters (ENSO), with an irregular period of about 3–7 years. The third graph shows well-known cycles of 11-year period attributed to the activity of the surface of the sun. Finally, the fourth graph shows the anomalies observed in the few large volcanic eruptions (earthobservatory.nasa.gov/Features/Globalwarming/page3.php)

2.3   The Feedback Concept

The concept of the feedback is very important in each dynamic model of systems due to the fact that it is the element that is capable of causing unexpected responses and, at the same time, of accounting for more complex behaviours of the system. Figure 2.4 shows the block diagram of the feedback. In the absence of the feedback block B, the block F processes the input quantity and converts it into a different output quantity O.

For example, imagine that the input variable I is the atmospheric concentration of CO_2. An increase of such a concentration entails, for the greenhouse effect, an increase of the earth's surface temperature that, in turn, causes the melting of a certain area of the permafrost at high latitudes. The *transfer function F* is a mathematical function that transforms a given change in the CO_2 input variable in a given variation of the output quantity A_p (area of melted permafrost) through a series of parameters set in the system. In our example, the parameters could be: (a) a rise in temperature caused by a unit increase of the CO_2 concentration; (b) the permafrost area a_p that melts for each degree increase in temperature T. This for that which regards the direct chain F, but it is known that in permafrost ice, non-minor amounts of greenhouse gases are trapped, among which carbon dioxide that is added to that which was already growing as a system input. The function that links the area of permafrost that melts to the amount of CO_2 that is released is represented by the transfer function B with the associated emission parameter of carbon dioxide per m^2 of melted permafrost. With this solution, the amount input to the F direct block is no longer I but I + BO. It is easy to show that the transfer function of the whole block, that is to say, the direct chain and the chain feedback is (Åström and Murray 2010):

$$G = \frac{F}{1 \pm FG}$$

If the effect of the feedback loop is added to the variation at the entrance of the direct chain, this is a *positive feedback* and the + signs matter, both in the block diagram as well as in the formula. If, instead, the effect of the feedback is subtracted from the change in input, this is called a *negative feedback* or *stabilizer*. The

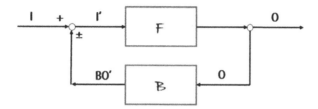

Fig. 2.4 Partial block diagram of a feedback system. In this scheme, a part of the output signal back to entrance treated by the feedback block. The composition between the variation of the signal I and the feedback signal BO', determines the stability or instability of the whole system

beneficial effect of the negative feedback is embodied in reporting, within certain limits, the output variable at the level it had before the perturbation. The positive feedback, however, has a generally destabilizing effect and can lead to the loss of control on the level of the output variable, triggering very often, considerable and undesired fluctuations. In the worst cases, it can lead to the collapse of one or more variables with tipping points.

In fact, the block diagram of the system should be further modified by introducing, in both the direct chain as well as the feedback chain, delay blocks to better represent the overall effect of the response of the system itself. The modelling of the feedback cycles is crucial for predicting the behaviour of a complex system such as the one formed by the continental surface, the oceans and the atmosphere of our planet. At present, an intermediate complexity model of the Earth system envisages several hundreds or thousands of direct and indirect feedback cycles. This makes the evaluation of the response to one or more radiant forcings extremely complex, so much so that they are natural, being of an anthropic nature. In these cases, the models in stock and flows play a crucial role in the construction of scenarios based on behavioural assumptions of external agents (almost exclusively solar irradiation) and of those that are anthropogenic. It is beyond the scope of this brief introduction to completely describe the substantial difference that exists between *prediction* and *simulation* (from this latter descend the scenarios). That which is interesting to emphasize here is that the effect of radiative forcing (of any kind) must be assessed taking into account the influence of all possible feedback blocks distributed within the model, more or less simplified, of the Earth system.

2.4   Enhanced Greenhouse Effect

We have seen that in its long history, the Earth has seen the alternation of periods of the dominance of low temperatures (ice ages), followed by periods (interglacials) in which higher temperatures dominated. A careful observation of the first diagram in Fig. 2.3 shows that over the past 800,000 years[2], 10–11 glacial periods have followed and as many interglacial periods. One can easily deduce that the average duration of a complete cycle is about 80,000 years, which, very roughly, tells us that, *again on average*, half of the period sees the rise of temperatures and the other half sees the descent. In fact, we know that this is not exactly right being that, for example, the interglacial periods tend to last less than the ice ages but what interests us here is the order of magnitude and not the exactness. Therefore, even if it is

---

[2]The empirical evidence presented in Fig. 2.3 were obtained through observations in the polar ice cores and ocean sediments. It is, therefore, of indirect observations of the temperature. However, the vast majority of scholars believe these very reliable projections with a confidence interval is shown with shaded color in the same figure. Using the same techniques it was possible to measure the concentration of GHG in the atmosphere trapping in the deep ice of air particles of the various geological periods.

evident from the graph that the ascent times of temperature are much shorter than those of the descents, the rise process from their minimum to their maximum still requires several tens of thousands of years. It must be highlighted however, that, in any case, we are still speaking about very long periods.

The extreme irregularity of the cycles does not permit reliable predictions, in any way, in the short to mid-term, by traditional means of statistics, and this also applies to those that predict, in *relation to only the natural variations*, the return of a coming Ice Age, or the aggravation of rising temperatures in the current interglacial period. We will see below that the combination of a high number of phenomena and especially the simulations via the dynamics of complex systems permitting the drawing of some conclusions with a high degree of reliability. It is important to also highlight that the instrumental direct measurements date from the beginning of the eighteenth century and satellite measurements from the second half of the twentieth century.

That being stated, it is worth remembering that, in the absence of other non-natural forcing, the global surface temperature of the Earth, on having reached the thermodynamic equilibrium between incoming and outgoing energy from the system, would stabilize around 15 °C. But from the time when mankind abandoned hunting and farming as the main means of support, one can observe the beginning of a period for which it is necessary to introduce, between climate radiative forcing, even the action of mankind that is essentially expressed in two ways: through the emission of greenhouse gases in the atmosphere and through the modification of the physical characteristics of the solid ground.

To many it seems hardly credible that human presence is able to compete with the enormous forces of nature, yet Fig. 2.5 can be useful in giving a proportion of the physical framework within which the thermodynamic equilibrium of the whole Earth system can be determined. The average radius of the Earth, measured above sea level, is 6373 km, the average depth of the oceans is about 3.5 km and their deepest point is approximately 11 km. The upper limit of the troposphere (TOA) is located about 12 km above sea level: in this thickness all the most important meteorological phenomena occur, from evaporation to transpiration, from the formation of the clouds to precipitation. It is in this space that the bulk of the greenhouse gases accumulate and it is here that they exert their reflection action of infrared thermal energy emitted from the Earth's surface. From 12 km to about 45 km lies the stratosphere. At the boundary between the stratosphere and the mesosphere, ozone is mainly distributed (and its so-called "hole") that, as we have already seen, filters out ultraviolet solar radiation.

Limiting interest to the last 1000 years of history, one could think of "splicing" the paleoclimate temporal observations with the instrumental observations available from the mid-nineteenth century. Basically, this is the work proposed by Michael Mann, Raymond Bradley and Malcolm Hughes in 1999, which gave rise to the perhaps the most contested, graphic in the history of science: the so-called "hockey stick" graphic. The paper of Mann et al. (1999) concludes with observations that the twentieth century was the planet's warmest century of the last 1000–2000 years, which the warming was extraordinary after the First World

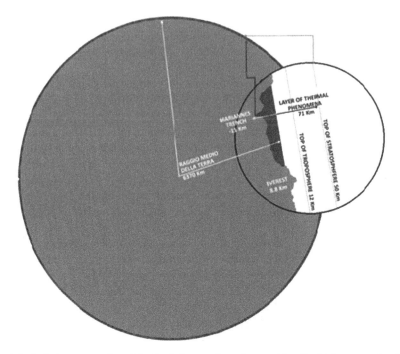

Fig. 2.5 Scaled representation of the Earth System. It can be noted that the thickness ranging from the deepest point of the oceans to the height of the stratosphere (*dark blue*) is 71 km away. Within this thickness occur all thermodynamic phenomena that lead to balance the thermal balance of the system itself

War, and finally that it reversed the natural cooling trend determined by the probable arrival of a new glacial cycle. The criticisms levelled at Mann's work concern both its operative aspects as well as those statistical and methodological. Since the publication of the chart to date, however, marked advances in climate palaeontology have been observed, with the availability of a large number of the surface temperature proxies ranging from the boring of the glaciers, to the examination of subsurface temperatures, from the analysis of stalagmites to that of wells, from the study of pine trunk sections to those of lacustrine sediments. All the techniques converge, in any case, towards the same conclusion: the twentieth century was the warmest of the last 1000 years and the mean global surface warming had a sharp rise after 1920 (Fig. 2.6).

However, only the increase in temperature is not sufficient to attribute the cause to human activity. For now, one can seek the necessary correlation with the greenhouse effect, and that is to say, with the variation of gas concentration that may have led to such an increase. In fact, as shown in Fig. 2.7, the set of measurements obtained from ice bores and those obtained from direct atmospheric measurements show a moderate trend of increasing concentrations of the very same concentrations between 1750 (the beginning of the industrial era) and 1950, followed by a distinct change of declivity after 1950. This direct correlation

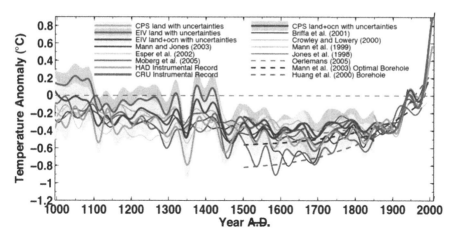

Fig. 2.6 Composite reconstruction of the Earth surface temperatures from 1000 to 2000 A.D., according to various methods and authors reported in the caption. The measures relate to the northern hemisphere, and for some series are shown with outline *gradient*, the confidence intervals. All projections converge on one conclusion: since mid-eighteenth century the temperature downward trend stops, until to reach, at the end of the twentieth century, temperatures never observed in previous years (1300–1700) (Mann et al. 2008)

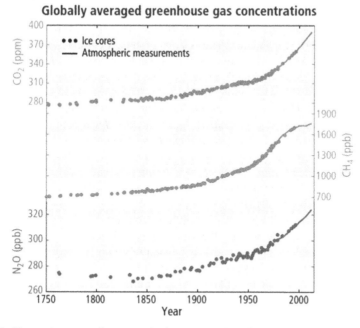

Fig. 2.7 Observations regarding atmospheric concentration of the main GHG: carbon dioxide (CO_2, *green*), methane (CH_4, *yellow*) and nitrous oxide (N_2O, *red*). The solid lines relate to the observations made directly into the atmosphere, while the points values are deducted from ice cores. (IPCC 2014)

between emissions and concentrations of greenhouse gases in the atmosphere and temperature increase is completed by the fact that, contrary to past ages, the thermal increases follow those of the concentrations of greenhouse gases and not vice versa. Despite this, the approaches of this type, typically regressive, are complicated by the superposition of the effects due to natural variations in solar radiant energy, to land-use changes (primarily variations in the earth's albedo), to the concentration changes in greenhouse gas of an anthropogenic origin (among these, the role played by aerosols) and, finally, the effects of the feedback.

To overcome these difficulties, it was decided to resort to the synergies between the regressive approaches and dynamic models of the climate. With climate models, in particular, it is possible to collect in a single stocks and flows system, the influences of each of the afore-mentioned effects with the additional advantage of using such models, once they have been duly tested, in order to separate the effects and to build scenarios by acting on a single variable, *equal to the behaviour of all the others*. The application of these models illustrates how the behaviour of the global average surface temperature of the Earth cannot be explained, in recent decades, only by natural phenomena without considering the contribution of human activity that proves crucial.

It is worth pointing out that the conflicts between scholars have developed in the past both on the reality of global warming, as well as in relation to the *attribution* of this warming to anthropogenic causes. At present, the improvement of observations, the most comprehensive analysis of paleoclimate evidence and the upgrading of models and the computational power of computers seem to favor, *in probabilistic terms*, toward the conclusion that there is a real problem of global warming and that it is, for the most part, due to human activities that have been developing since the beginning of the "industrial age" along with the demographic developments and phenomena linked to urbanization and land use. The terms used in the 5th IPCC report are clear: *"Human influence on the climate system is clear, and recent anthropogenic emissions of greenhouse gases are the highest in history. Recent climate changes have had widespread impacts on human and natural systems... Warming of the climate system is unequivocal, and since the 1950s, many of the observed changes are unprecedented over decades to millennia. The atmosphere and ocean have warmed, the amounts of snow and ice have diminished, and sea level has risen."* (IPCC 2013).

A more complete analysis of the scientific reasons for the contrasts between the two sides in the field (which, by the way, are significantly different as well as numerous), although of great interest, is beyond the scope of this text. What it is worth pointing out is that, beyond any dispute, it is necessary that humanity faces the consequences of a more than likely global warming following two paths: that of adaptation to the consequences of this phenomenon and that of its mitigation.

2.5 Adaptation and Mitigation

The quoted Synthesis Report of the fifth report of the Intergovernmental Panel on Climate Change (2014) reads: *"Effective decision-making to limit climate change and its effects can be informed by a wide range of analytical approaches for evaluating expected risks and benefits, recognizing the importance of governance, ethical dimensions, equity, value judgments, economic assessments and diverse perceptions and responses to risk and uncertainty"*.

It is a reminder that, whatever the causes are that drive climate change, they deserve to be taken seriously by governments. The already observed effects of global warming and those predicted for the future mainly regard the rise in sea level, more frequent floods and droughts, the greater frequency of heat waves, and the greater frequency of extreme events. Even if anthropogenic emissions of greenhouse gases were to be blocked as from tomorrow, we will still have to cope with these issues for several decades, given the long persistence of these gases in the atmosphere and given the enormous amount of thermal energy (heat) already stored in the oceans. In addition, one must consider that the climate models, due to the feedback and the delays distributed in the climate system, foresee the risk of a sudden and irreversible change with absolutely tragic consequences (*tipping point*).

The effects of the ongoing climate change are the most diverse and distributed differently along the globe. In general, it can be said that the most vulnerable populations, with a shortage of resources to organize a viable preventive defence against the expected impacts will be most affected. The most important of these expected impacts include: water scarcity; coastal erosion associated with floods and tsunamis; decreased food production in areas already heavily put under stress; increased mortality due to heatstroke; problems for fish production; damage by fire; movements towards the north of the area of residence of animal and plant species; strong risks for some ecosystems and species now already precarious.

These risks can be limited by adaptation actions whose implementation is already considered long overdue. The reluctance of the overwhelming majority of countries to plan and carry out the works and measures necessary for adaptation is mainly due to their scarce political return in the short term that today seems the main development of economic policies. Adaptation efforts, no longer postponable, require heavy investments whose real value can only be assessed with a horizon longer than that between one election and another. This circumstance, coupled with the continuing global economic crisis, makes the predictions about the impacts of climate change very worrying. From a political point of view, the temptation to implement interventions, only once a disaster has occurred, is very strong, due to the fact that, among other things, this approach makes it more acceptable to the population an increase in tax burden, (where possible), presented as an absolute necessity to cope with emergencies. In fact, it has been proven that the cost involved in repairing damage already incurred is far higher than that which would be faced with a more far-sighted prevention policy.

In terms of mitigation, however, something started to move in 1992 as part of the United Nations Framework Convention on Climate Change (UNFCCC), which, since the Conference in Rio de Janeiro, placed emphasis on the need to reduce emissions of greenhouse gases, with particular regard to carbon dioxide. Since the first conference in Rio, 22 Conference of the Parties have taken place (to say nothing of the parallel conferences) from 1995 (in Berlin), to 2016 (in Marrakech), an almost continuous debate without which, however, a real reduction of emissions has been achieved nor a draft of binding regulations (including penalties) for the foreseeable future. On the other hand, a series of meetings to monitor concentration levels of greenhouse gases with the underlying objective of limiting the increase in global surface temperature in the range of 2 °C compared to the pre -industrial age and by the end of this century, have taken place. This would require, according to climate models, the reduction of about 70% of CO_2 emissions. An ambitious goal that, at the same time, is judged to be insufficient by the majority of climate scientists.

Achieving the goal requires: (1) the substantial replacement of fossil fuels with renewable energy sources; (2) greater efficiency in both the production and consumption of energy; (3) a slowdown in global economic growth; (4) a reduction of the population. To these, one should also add that which is related to the capturing and stocking of carbon. Naturally, these intervention measures separate the countries that should apply them, both for the associated economic costs as well as with regards to the social aspects that would ensue. It must also be taken into account that direct policies to encourage the development of clean technologies, applied to date, have yielded results that are not entirely satisfactory. Even those who believe that the problem can be solved by recourse to a compulsory carbon tax, do not take into account the effects this would have on those economies that are starting their development based not only on low labour cost, but also on scant attention dedicated to environmental problems whose solution is still considered a luxury.

An easier way forward would be to accompany and promote the development of poor countries and emerging markets through investment flows that start from the developed countries, which have been and still are responsible for the majority of CO_2 emissions. These flows are aimed at the mitigation of emissions and the adaptation of the said countries to the effects of environmental changes. The Finance of Climate policy promises to be among the most effective tools for mitigation as it would facilitate, at the same time, both the development and sustainability of the most disadvantaged countries.

References

Åström KJ, Murray RM (2010) "§1.1: What is feedback?". Feedback Systems: An Introduction for Scientists and Engineers. Princeton University Press/Cambridge University Press, Cambridge/New York, p 1

Hansen J, Ruedy R, Sato M, Lo K (2010) Global surface temperature change. Rev Geophys 48(4)
    IPCC, 2013: Climate change 2013: The physical science basis. Contribution of working group I
    to the fifth assessment report of the intergovernmental panel on climate change (Stoker TF,
    Qin D, Plattner G-K, Tignor M, Allen SK, Boschung J, Nauels A, Xia Y, Bex V, Midgley PM
    (eds.))
Mann ME, Bradley RS, Hughes MK (1999) Northern hemisphere temperatures during the past
    millennium: Inferences, uncertainties, and limitations. Geophys Res Lett 26(6):759–762.
    doi:10.1029/1999GL900070
Mann ME, Zhang Z, Hughes MK, Bradley RS, Miller SK, Rutherford S, Ni F (2008) Proxy-based
    reconstructions of hemispheric and global surface temperature variations over the past two
    millennia. Proc Natl Acad Sci USA, 105(36): 13252–13257, doi:10.1073/pnas.0805721105,
    PMC 2527990, PMID 18765811
Trenberth K, Fasullo J, Kiehl J (2009) Earth's global energy budget. Bulletin of the American
    Meteorological Society
IPCC (2013) Climate change 2013: the physical science basis. Intergovernmental Panel on
    Climate Change Report Overview
IPCC (2014) Climate change 2014: fifth assessment report. Intergovernmental Panel on Climate
    Change Report Overview

Chapter 3
Climate Finance

Abstract Since 1992, the United Nations Framework Convention on Climate Change (UNFCCC) has set out a framework for international action to stabilize greenhouse gas emissions (GHG) to prevent dangerous climate change. The UNFCCC recognizes that developed country Parties (Annex II Parties) have to provide financial resources to assist developing country Parties in implementing the set goals.

During the 15th session of the Conference of the Parties (COP15) held in Copenhagen, the developed countries pledged to cooperatively provide USD 100 billion annually by 2020 to developing countries and guaranteed immediate "Fast-start Finance" of up to USD 30 billion over 2010–2012 to launch the project. Such financing can support developing nations by enabling poorer countries to move towards low-emission and climate-resilient development pathways.

In this chapter we focus on climate finance as it appears from the meetings that have taken place from the COP1 to the COP22. We propose a first analysis of the financial flows intended to promote energy generation and supply with renewable sources and those funds intended for the protection of the physical environment in order to identify preferential channels in "Fast-start finance" distribution and whether these preferences can affect the effectiveness of such measures.

Keywords Conference of the Parties • Climate finance • Fast Start Finance • Funds' distribution • Energy generation • Biosphere protection

3.1 From Rio de Janeiro to Marrakech: a Brief Summary

When the international community, at the level of governments, began to become aware of the unalterable nature to combine general economic development to its sustainability from an environmental point of view, it felt it had to schedule a series of meetings between the representatives of the member countries to deal systematically with the problem of global warming and the mitigation of chemical and physical agents that caused it. The first meeting was held in Rio de Janeiro between 3rd and 14th June 1992 and was attended by 172 countries and 2400 representatives of non-governmental organizations. This meeting produced an Agenda for the XXI

© The Author(s) 2018   23
A.A. Romano et al., *Climate Finance as an Instrument to Promote the Green Growth in Developing Countries*, SpringerBriefs in Climate Studies,
DOI 10.1007/978-3-319-60711-5_3

century, a Convention on Biological Diversity, a General Declaration and a Convention on Climate Change. On the occasion of the meeting (known as the Earth Summit) a Commission on Sustainable Development responsible for assisting and monitoring the implementation of Agenda XXI, was set up. The flaw in the Meeting is that the treaty that resulted did not set out obligatory constraints to emissions or provide for any form of sanction albeit contemplating the possibility of adopting future protocols with mandatory limits. On the occasion of the Conference, the clustering of countries into three groups was utilised: those industrialized, those industrialized that allocate financial flows to developing countries and countries in the developing world. This division into groups was and still is widely used in scientific literature.

The UNFCCC in Rio de Janiero was followed, according to the agreements, by a series of "Conferences of the Parties" (COP) that occurred over time until the last in 2016 in Marrakech. In each of these conferences topics were discussed and regulatory decisions were taken that can be summarized below:

- *COP1 – Berlin, 1995*. Here emerged the awareness of the difficulties and the resistance from some states to fulfil their "obligations" set into place by the Conference. It is here that began the "national ways" to compliance with exemptions established for countries *Not in Annex I* even if these were included among the countries that would be (and indeed are) among the most polluting.
- *COP2 – Geneva, 1996*. The second IPCC Report (AR2) of which the Conference took note, had been published the year before. Officially it accepted the policy of "flexibility" advocated by the United States with regard to the actions to be carried out and manifested the need to issue binding regulations for the foreseeable future.
- *COP3 – Kyoto, 1997*. Of all the conferences, this is the best known to the public. Its fame was largely due to the political tensions between the participating countries that characterized it, and the resulting interest by the worldwide media. The Conference produced an international treaty known as the Kyoto Protocol which has become a reference point for all subsequent actions. The protocol envisaged for the period 2008–2012, a reduction of greenhouse gas emissions of a percentage not less than 8.65% compared with 1985, that was established as the base year. The identified greenhouse gases are carbon dioxide (CO_2), methane (CH_4), nitrous oxide (N_2O), hydrofluorocarbons (HFCs), perfluorocarbons (PFCs) and sulphur hexafluoride (SF_6). In particular, the United States would have had to reduce their emissions by 7% compared to 1990. The meeting also envisaged some indirect mechanisms to reduce emissions through which the industrialized and emerging countries could acquire emission credits through the implementation of projects for the purpose in developing countries. In addition the so-called *Emission Trading*, that is, certificates that could be issued and traded by performing countries that met and exceeded the targets assigned to them, were also created.
- *COP4 – Buenos Aires, 1998*. Two years after the implementation period of the Kyoto Protocol, some problems that remained unresolved from the previous

conference, should have been resolved. Instead irreconcilable divisions emerged that led to a mere "Declaration of Intent" to be implemented in the subsequent years.

- *COP5 – Bonn, 1999.* It was merely a technical meeting that bore no significant results.
- *COP6 – Aja, 2000.* This conference, which began with ambitious goals, recorded a substantial break between the United States, the United Kingdom and the European Union. Once again, there was hostility by some countries towards the fulfilment of previous commitments. The work was suspended and it was decided to continue the conference in Bonn as a continuation of the current meeting.
- *COP6 B, – Bonn, 2001.* On this occasion, the United States decided to participate in the Conference as observers only after they had refused the Kyoto agreements. Despite the bad omens, the Conference was able to find some important agreements, among these: the revival of the Clean Development Mechanism (DCM) with the elimination of limits on the credit gained by developed countries thanks to investments in Developing Countries; the reduction of carbon into the atmosphere through forestation and reforestation actions by countries that with this mechanism would acquire credits; some small mention of sanctions (the freezing of credits earned) for failure to achieve the targets set for each country; three new funding for the implementation of mitigation actions.
- *COP7 – Marrakech, 2001.* Here the date of 2002 for the entry into force of the Protocol, was established. In addition, some operational measures were taken and actions were undertaken to enlarge the number of countries acceding to the Protocol.
- *COP8 – Milan, 2003.* The Conference decided to use the funding already allocated to Marrakech to support projects in developing countries for measures to adapt to the effects of climate change and the transfer of technology to these countries.
- *COP10 – Buenos Aires, 2004.* No significant decisions were taken.
- *COP11 – Montreal, 2005.* This Conference devoted much of its time to the discussion on the need to drastically reduce emissions of chlorofluorocarbons (CFCs), responsible for the so-called "ozone hole" which would lose much of its ability to filter ultraviolet radiation with its danger to human health.
- *COP12 – Nairobi, 2006.* The Conference made some progress regarding the emission reduction plan after 2012 and decided to involve some African states in the clean development mechanisms.
- *COP13 – Bali, 2007.* No significant decision was taken.
- *COP14 – Poznan, 2008.* No significant decision was taken.
- *COP15 – Copenhagen, 2009.* Various proposals to cut CO_2 emissions without any mention of constraints and sanctions were established. At the end of the meeting, when the Conference seemed to be heading towards a failure, it reached an agreement of a limit of 2 °C of global warming, it established a budget of 30 billion dollars to curb the effects of climate change in developing countries

and a further 100 billion dollars to be allocated to the weaker countries to promote economic development through clean energy. It was the first base of *Climate Finance.*

- *COP16 – Cancun 2010.* With the Cancun Conference some steps forward but essentially qualitative rather than quantitative, were made. The formal decisions made by the participants restarted the discussion on the reduction of greenhouse gas emissions and outlined a number of measures for adaptation to global warming. After the failures of the previous conferences, the Cancun agreement awakened interest and hopes.

- *COP17 – Durban, 2011.* The Conference reaffirms the central role of the Kyoto Protocol envisaging a second period of legal and financial commitment. There was the search for a new definition process capable of leading, in the future, to binding decisions on behalf of the member countries. Finally, the decision to make the Green Fund (GCF Green Climate Fund-US $ 100 billion by 2020) operational from 2012, was taken.

- *COP18 – Doha, 2012.* It was a very complex Conference, which took place around a large number of tables from which it is difficult to draw a valid conclusion other than the fact that negotiations could still continue.

- *COP19 Warsaw, 2013–* Once again there was a confrontation on the historical responsibility for global warming between the emerging countries, China and India, and the United States, which threatened to derail the conference. To this one can add the protest of the developing countries who complained about the non-receipt of the financial assistance planned in the past. As often happened at the Conferences of the Parties, a small compromise on strictly formal plans, was finally reached.

- *COP20 – Lima, 2014.* It was a substantial failure with a strong protest made due to the presence of Multinational Companies at the Conference. In the general confusion, a further formal compromise in quality was reached.

- *COP21 – Paris, 2015.* Great hopes were pinned on this Conference and it concluded, for the first time, with the general agreement of all 196 participating countries. The limit set for the entry into force of the agreement is 2020, and this attracted significant criticism from various sectors. The most important result was to restrict "as soon as possible" carbon dioxide emissions and limit global warming to "below" 2 °C (preferably 1.5 °C) compared to the level of pre-industrial times; controls every 5 years were planned with a commitment to arrive as early as 2018 with substantial cuts in emissions; the old industrialized countries pledged to finance with $100 billion a year the dissemination of low-carbon technologies; the implementation of a reimbursement mechanism for damage from global warming to the most exposed countries, was also planned. Many critics pointed out the deferred start, the lack of a date for the elimination of emissions, the self-certification of controls, the powers conferred to the oil barons and, once again, there was the absence of sanctioning rules for defaulting countries. It must be recognized, however, that for the first time, the Conference unanimously admitted that "Climate change represents an urgent and potentially irreversible threat to human societies and the planet".

- *COP22 – Marrakech, 2016.* No major decisions other than those to postpone to 2018 the date for the reduction of emissions were taken. It is necessary to highlight the protests of the poorest countries for the non-payment of 100 billion provided for by the Paris Agreement. The agreement to fix the institutional rules and the determination of the operation for the disbursement of funds for adaptation was postponed to the end of March, 2017.

3.2 Fast Start Finance

Since 1992, the United Nations Framework Convention on Climate Change (UNFCCC) has set out a framework for international action to stabilize greenhouse gas emissions (GHG) to prevent dangerous climate change. The UNFCCC recognizes that developed countries have contributed the most to the global accumulation of GHG emissions, while developing countries bear less historical responsibility. This recognition has led to a commitment from developed countries to mobilize finance to help developing countries respond to climate change, and such 'climate finance' has become a central issue in international negotiations.

During the 15th session of the Conference of the Parties (COP15) held in Copenhagen, the developed countries pledged to cooperatively provide USD 100 billion annually by 2020 to developing countries and guaranteed immediate "Fast-start Finance" of up to USD 30 billion over 2010–2012 to launch the project (UNFCCC 2014). Such financing can support developing nations by enabling poorer countries to move towards low-emission and climate-resilient development pathways.[1] Particular attention was dedicated to vulnerable countries, including Small-Island Developing States (SIDSs), Least-Developed Countries (LDCs), and African States. For these reasons, during COP15 to the UNFCCC the parties agreed to create the Green Climate Fund (GCF) as a new operating entity of the financial mechanism for the UNFCCC. The GCF was effectively launched at COP17 (held in Durban in 2011). The new organization is the main channel through which climate finance is allocated. The GCF, which is headquartered in South Korea, is controlled by a Board on which developed (Box 3.1) and developing (Box 3.2) countries are equally represented (Ellis et al. 2013). To combat the effects of climate change and the increasingly frequent demand for natural resources, the international community recently signed a climate agreement during the 21st session of the Conference of the Parties held from 30 November to 12 December 2015 in Paris. This effort has resulted in the adoption of the first international agreement to limit the increase in global temperature that takes into account *(i)* adaptation, i.e., actions taken to help

---

[1]Global total climate finance directed to developing countries for mitigation and adaptation activities has significantly increased (by nearly 15%) since 2011–2012 (UNFCCC 2016b). In a status report on the amount of funding mobilized by developed countries (2015), the OECD argued that there has been significant progress and that such funding reached USD 52 billion in 2013 and USD 62 billion in 2014.

communities and ecosystems manage changing climate conditions, and (*ii*) mitigation, which aims to constantly reduce GHG emissions (UNFCCC 2016b).

Developing country governments are rightly concerned about potential tensions between promoting the economic growth needed to generate jobs and reduce poverty, and reducing anthropogenic impact, in particular the GHG. International cooperation on finance has the potential to help countries manage such trade-offs, and create new incentives for low carbon development. With the Fast-start finance the Annex-II Parties of the Kyoto Protocol are required to provide financial resources to enable developing countries (Non-Annex-I) to undertake emissions reduction activities and to help them adapt to the adverse effects of climate change.

International cooperation on finance has the potential to help countries manage such trade-offs and create new incentives for low carbon development. Climate finance can support the policies that can build resilience against the threats posed by climate change (Nakhooda and Norman 2014).

The global climate finance architecture is complex: finance is increasingly channeled through bilateral channels as well as through multilateral funds – such as the Global Environment Facility (GEF), the Climate Investment Funds (CIFs), and the Adaptation Fund (AF). In addition, a growing number of recipient countries have set up national climate change funds that receive funding from multiple developed countries in an effort to coordinate and align donor interests with national priorities.

Box 3.1 Developed Countries Definition

For the scope of this book we have adopted the definition of developed countries proposed by UNFCCC. This includes the 20 UNFCCC Annex II parties that provide climate funding to recipient countries. Annex-II Parties consist of the OECD members of Annex-I but not the economies in transition (EIT) Parties. Annex-II Parties are required to provide financial resources to enable developing countries to undertake emissions reduction activities. EIT include the Russian Federation, the Baltic States, and several Central and Eastern European States.

Hereafter referred to collectively as "developed countries" or "donor countries".

Box 3.2 Developing Countries Definition

For the purposes of this book we consider as developing the countries that receive climate funding from donor countries. This generally comprises countries included in Non-Annex I Party. Non-Annex-I Parties (including the least developed countries – LDC), are mostly developing countries. Certain groups of developing countries are recognized by the UNFCCC as being especially vulnerable to the adverse impacts of climate change.

There is generally much more transparency about the status of implementation of multilateral climate finance initiatives than of bilateral climate finance initiatives. However, only 18% of the Fast Start Finance mobilized between 2010 and 2012 was spent through multilateral climate funds with 62% spent bilaterally (Nakhooda et al. 2013). The proliferation of climate finance mechanisms increases the challenges of coordinating and accessing finance.

The number of multilateral implementing agencies has expanded from the three original founding partners of the GEF i.e. the World Bank, the United Nations Development Programme (UNDP) and the United Nations Environment Programme (UNEP), to include more than 40 institutions (see, e.g., Nakhooda and Norman 2014). This expansion results, in great part, from innovations introduced through the Adaptation Fund that facilitated developing country-based institutions to have direct access to climate finance. The range of partners for climate funds now includes regional development banks, a range of international organizations, developing country ministries, trust funds and non- governmental organizations (NGOs).

3.3   The Framework to Assess the Climate Action

The research about the climate finance and its main drivers, i.e. the availability of funds and the disbursement mechanisms, are important elements for assessing effectiveness of financed programs in reaching their objectives.

Track the flow of climate finance received from international sources at the national level and analyze the effectiveness of these flow of funds to promote the green growth in developing countries is a complex issue that involves a multitude of stakeholders and requires a careful analysis of the data sources.

In a report presented during the 21st COP in Paris, the OECD, in collaboration with Climate Policy Initiatives (2015), provided a status check on the level of climate finance mobilized by developed countries in 2013 and 2014. The preliminary estimates provided in this report are that climate finance reached USD 62 billion in 2014 and USD 52 billion in 2013, equivalent to an annual average over the 2 years of USD 57 billion.

Nevertheless the preliminary estimates presented, in the report they highlights how "recent developments in definitions and accounting methodologies to track climate finance are a staging post on the way towards more complete and transparent estimates of climate finance" and "further analytical and methodological effort will be required to underpin future improvements in measuring and reporting climate finance across a range of organizations, international financial institutions and countries".

Despite the difficulties for the quantification of the funds, many authors have debated about climate finance and the potential of international cooperation to promote economic growth and reduce its environmental impact by promoting green electricity generation.

Nakhooda and Norman (2014) maintain that without adequate financial support, i.e., one that bears the costs of renewable energy sources (RES) deployment, the governments of developing countries must choose between economic growth (to generate jobs and reduce poverty) and climate protection. The authors argued that financial cooperation among the developed countries has the potential to help poor countries by creating new incentives for low-carbon development and to support policies to build resilience against the risks caused by climate change. Keeley (2016) focuses on the importance of attracting funding from donors to develop resources, given the importance of international financial recourses for many developing countries or those that are heavily dependent on fossil fuels and that are subject to significant fluctuations in oil prices. Espagne (2016) focused their attention on the difficulty that developing countries face in developing their economic systems without using carbon fossil sources and relying on the financial support of developed countries. Bazialian et al. (2011) and Bhattacharyya (2013) argued that access to electricity remains a key question for several developing countries. In particular, Bazilian et al. (2011) consider poverty one of the primary obstacles must be overcome to drastically improve the investment for energy access. These research conclusions are even more remarkable if we consider that a country's economic development depends on how aid is organized and distributed to develop a sustainable system of energy production and to optimize the transmission and distribution electricity grid (which is mostly lacking in the developing countries).

Furthermore, Marquardt et al. (2016), in a study of financial aid for the transition to an energy-efficient and low-carbon economy in the Philippines and in Morocco, argue that while climate finance cannot be imposed on these countries, it may be necessary to lead them through a process of gradual change – through niche level projects – to test the effects of changes.

Regarding the management of financial aid, several authors (Bird et al. 2013; Urpelainen 2012; Bourguignon and Platteau 2015) have emphasized the importance of intra-governmental coordination by proposing a balanced, supranational organization to manage climate funding and thus avoid the expensive duplication of projects. In addition, Bird et al. (2013) describe an approach to measuring the effectiveness of the national systems that support climate finance delivery. They assess three interlinked elements of government administration: the policy environment that supports climate change expenditures, the institutional architecture that determines relevant roles and responsibilities, and the public financial system through which climate change-related expenditures are channeled.

Similarly, Ellis et al. (2013) and Tirpak et al. (2014) investigated how climate finance effectiveness is monitored and evaluated in different communities. Ellis et al. (2013) explore how different communities view climate finance effectiveness, the policies or institutional pre-conditions that facilitate effectiveness, and how effectiveness is currently monitored and evaluated. Tirpak et al. (2014) present nine technical, political, and capacity challenges faced by developing countries that were discussed during three workshops in Asia, Africa, and Latin America. Participants in these workshops discussed some of the steps that developing countries and their international partners can take toward monitoring and tracking climate finance more effectively.

Pickering et al. (2015) investigate the intra-governmental dynamics of climate finance decision-making in some donor countries (Australia, Denmark, Germany, Japan, Switzerland, the United Kingdom and the United States). They highlight the importance of intra-governmental coordination in the management of climate aid. Bigsten and Tengstam (2015) seek to quantify the effects of improved donor coordination on aid effectiveness. They find that aid coordination efforts may reduce donor transaction costs and increase the possibilities of achieving donor objectives in recipient countries, but there will also be political costs to the extent that the donor loses some political control over aid transfers.

Despite several seminal papers on climate finance, the assessment of the effectiveness of green funds in reducing GHG emissions and the effect of the funding concentration are topics that have not been adequately examined. To help policymakers address, plan and better target climate funding for long-term projects, it is important to fill this research gap.

3.4 Analysis of Flow of Funds from Donors to Recipients Countries

The role of climate finance is to guarantee a sustainable economic growth improving all aspects of the relationship between energy supply and environmental issue. To reach this goal the developed countries pledge to financing and support the developing pathways to safeguard the environment.

The assessment of the effectiveness of climate finance in reducing the impact of climate change and in promoting the economic growth is a relevant research topic because it can support policymakers in order to plan climate funds for long-term projects. The findings can propose areas for discussion in order to manage inter and intra governmental strategy in view of work program to be undertaken to implement the Paris agreement.

Developed countries, as such, committed to supply new and additional resources, including forestry and investments, during the COP15 in 2009. These resources amount to at least USD30 billion for the period 2010–2012, known as "Fast-start Finance", to be fairly distributed according to the mitigation and adaptation theme. On 12 December 2015, with the Paris Agreement (COP 21), all of the 195 UNFCCC participating member states and the European Union renewed their agreement to reduce GHG emissions, as well as their commitments and obligations.

It is against this complex and evolving backdrop that political and institutional relationships become very important, as assessed by Brunner and Enting (2014) and Pickering et al. (2015). The latter, in fact, argue for collaboration between development and environmental agencies to manage the funds of climate finance and to prevent their different missions from giving rise to disagreements about priority

Box 3.3 ODA and OOF Definitions

The OECD Development Assistance Committee (DAC) qualify as "official development assistance" (ODA) a flow that meet 4 criterion: (*i*) come from an official source, (*ii*) is focused on economic development and welfare of developing countries, (*iii*) have concessional terms and (*iv*) is destined to an ODA eligible country. Whereas, "other official flows" (OOF) are defined as official sector transactions that do not meet official development assistance (ODA) criteria (i.e., grants to developing countries for representational or essentially commercial purposes or other official transactions, that are primarily export-facilitating in purpose)

targets. Indeed, even though climate funds can be included in calculations as aid for Official Development Assistance (ODA) (see Box 3.3), the nature of the ODA funds differs from the aid directed to environmental protection, and the general objectives of the two types of funds may diverge (see, e.g., Pickering et al. 2015). As highlighted by Nakhooda (2013) and recently by Tol (2017), the disparities among the governments' priorities between environmental issue and a rapidly economic growth would make any opportunity of expansion of climate policy unsuccessful.

To better explain the intensity and the direction of the flow of funds we separately analyze (*i*) consistency and difference between committed and disbursed funds, and (*ii*) the preferential channels between donors and recipients countries. Moreover, we analyze how the received funds combat the climate change promoting the green growth.

3.4.1   Commitment vs. Disbursement

We want investigate the climate funds of "Fast-Start finance" focusing on bilateral flows between donors and recipients, to assess the effectiveness[2] of climate funding and prove the existence of preferential channels between developing countries. Thus, we analyze the flow of funds, highlighting the direction, the intensity and the environmental performance of the recipient countries. We report the flows of "Fast-start Finance" for the year 2010 (the data was retrieved from AidData.org, see Boxes 3.4 and 3.5 for details on project).

---

[2]By aid effectiveness, we mean ensuring that aid reaches the developing countries, which, due to their rapid growth, have become the main emitters of greenhouse gases.

Box 3.4 Commitment and Disbursement Definitions

Total commitments per year relate to pledges in the year in question and additions to agreements made in earlier years. Whereas the disbursements are the amounts of resources at the disposal of a recipient country or agency and it can take several years to disburse a commitment.

Thus, while commitments are tied to projects formed in a specific year, disbursements can be tied also to projects originating in previous years.

Box 3.5 AidData Project

In order to assess the effectiveness of "Fast-start Finance", commitments alone are not sufficient. For this reason, the chosen dataset refers to the AidData Research Release 2.1 (of which the last year is 2010) because the disbursements are not made available in AidDataCore_ResearchRelease_Level1_v3.0 (Released: 2016/04/29), which is mainly a commitment database (Tierney et al. 2011).

The dataset tracks known international development flows from bilateral and multilateral aid organizations. It also includes all forms of development finance, including Official Development Assistance (ODA), Other Official Flows (OOF), Export Credits, and Equity Investments (see Box 3.3 for details).

AidData was created from the merger of two prior initiatives, Project-Level Aid and Accessible Information on Development Activities. In 2010, it was released the first database with one million past and present development finance activities from over 90 funding agencies retrieved from several sources. This includes the OECD's Creditor Reporting System, annual reports and project documents published by donors, web-accessible database and project documents, and data obtained directly from donor agencies.

The research analyzes the flows destined to (*i*) energy generation and supply with regard to power generation by renewable sources and the flows of funds targeted to (*ii*) biosphere protection.

Furthermore, we consider the flows of funds targeted at (*i*) the general environment and (*ii*) biosphere protection (air pollution control, ozone layer preservation and marine pollution control). In general, both types of financial flows are among the aid directed to climate change, and these funds are for both climate adaptation and climate mitigation. These flows of funds represent a contributing factor to investments in alternative green energy usage, as they compensate for credit and liquidity constraints and improve technological progress in environmental protection (Zhao et al. 2012).

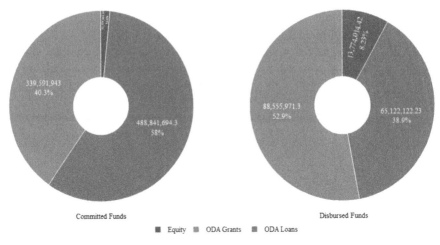

Fig. 3.1 Committed and disbursed funding in terms of financing type
Source: Our elaboration on AidData Database (AidData.org)

For that involving biosphere protection, aid are focused on activities that involve specific measures to protect or enhance the physical environment, without sector specification.

The data mainly relate to bilateral disbursed flows in 2010. However, to provide a clear and complete representation of the comparison between committed and disbursed funds, it was also retrieved the data related to projects where the flow was disbursed in 2010, but was originated in previous years. Thus, our analysis relate to 512 projects (7 of which related to period between 2007 and 2009), and about 16% of the analyzed projects (84 projects) did not have disbursed amounts (Fig. 3.1).

Donors provided to climate funds in the form of grants, capital contributions, or concessional loans. The largest form for committed funds is the ODA loans. With this form, despite to grants that includes the monetary value of in-kind contributions provided by donors and debt forgiveness, the provided projects must be paid back. However, both ODA loans and ODA grants are international flows focused on development (see Box 3.6 for details). The second largest form in the committed aid is ODA grants. While in the disbursed funding, the pattern is reversed.

The percentage of equity investment rises in disbursed flow and the only donor that has implemented this strategy was United Kingdom (UK), by Climate Public Private Partnership (CP3, see Box 3.7).

Another difference is in level of income of recipients (see Fig. 3.2). The funding should promote a reduction of GHG emissions and sustainable economic growth. The largest recipients are among countries within the classification of Upper middle income[3] with 53.3% of total commitments and 48.6% of disbursements (with Turkey receiving around 17.4% of commitment). It is followed by the countries belonging to Lower middle-income group. Our analysis also shows that there are

---

[3]The grouping follows the World Bank's classification by income (see Box 3.8 for details).

Box 3.6 Flow Type Definitions

Grant: A project where flows given do not need to be paid back. This can include the monetary value of in-kind contributions provided by donors. According to CRS directives, grants are transfers in cash or in kind for which no legal debt is incurred by the recipient.

ODA Grants: Grants that qualify as Official Development Assistance (see Box 3.1 for ODA definition)

Loan: A project where the funds provided must be paid back. If it is, "non-concessional" must be paid back using market interest rates or the terms less than 25% concessional while, if it is "concessional" the loan uses concessional terms.

ODA Loans: Concessional loans that qualify as Official Development Assistance (see Box 3.3 for ODA definition)

Equity: Some development financiers invest in a recipient country's institutions/companies. They purchase shares/equity in the company.

Export Credits: Loans for the purpose of trade and which are not represented by a negotiable instrument.

OOF Loans (non-export credit): includes official sector loans, which do not meet the ODA criteria (see Box 3.3 for OOF definition).

Box 3.7 Climate Public Private Partnership (CP3)

The aim of Climate Public Private Partnership (CP3) program is to prove that climate investments in developing countries are financially feasible.

The UK works with the Asian Development Bank and International Finance Corporation in a joint exertion to encourage capital funding into low carbon climate resilient investments.

The CP3 program invest directly in projects, with the form of equity investment, to support investments in energy efficiency, renewable energy and clean technology in developing countries. The equity investments, despite the grants, does not limit its distribution options and the activities it can support (ICAI 2014).

some countries belonging to High income, even if they receive a very small amount (Chile receive the 0.57% of total disbursement). As also assessed in Nakhooda and Norman (2014), the distribution of allocation should reflect country exertions to access the funding for which they may be eligible. The report shows also as wealthy developing countries have received small funding despite high GHG emissions (e.g., Iran, Saudi Arabia, and South Korea).

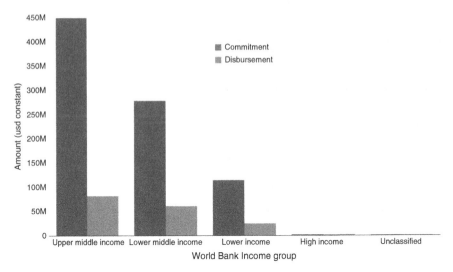

Fig. 3.2 Committed and disbursed funding following the World Bank's income classification of recipient countries
Source: Our elaboration on AidData Database (AidData.org)

However, the same report is regarding all mitigation allocations including the multilateral funding, while in our analysis we consider only bilateral aid to uncover preferential channel between donors and recipients. Our analysis, instead, focuses on only bilateral funding and only two type of aid, which are targeted to biosphere protection and power energy by renewable sources.

To explore the pledges of donors, a radar chart was created. The indicators of incidence of disbursement (as ratio of disbursed funding of committed fund) reported in Fig. 3.3, were arranged anti-clockwise from highest to lowest amount of commitment fund.

The Germany and United States have been the largest contributors. Other major contributor include France, Switzerland, Belgium and Japan.

The first notable fact about the funds' distribution relies on the major pledges made by Germany for which the effective disbursement was very small. It is followed by the United States that committed only 23% of Germany's pledges and, at the same manner, its disbursement was very small.

Fig. 3.3 Incidence of disbursement on pledges
Source: Our elaboration on AidData Database (AidData.org)

Box 3.8 Income Groups
In the World Development Indicators database (and most other time series
datasets), all 189 World Bank member countries, plus 28 other economies are
classified by geographic region, by income group, and by the operational
lending categories of the World Bank Group. Economies are currently
divided, by income, into four groups: low, lower-middle, upper-middle, and
high. Income is measured using gross national income (GNI) per capita, in
U.S. dollars, converted from local currency using the World Bank Atlas

(continued)

Box 3.8 (continued)

method. Estimates of GNI are obtained from economists in World Bank country units; and the size of the population is estimated by World Bank demographers from a variety of sources, including the UN's biennial World Population Prospects. According with the World Bank classification: (*i*) low-income economies are defined as those with a GNI per capita of $1025 or less; (*ii*) Lower middle-income economies that are those with a GNI per capita between $1026 and $4035; (*iii*) Upper middle-income economies are those with a GNI per capita between $4036 and $12,475 and (*iv*) high-income economies with a GNI per capita of $12,476 or more.

On the other hand, countries as Greece, Korea and Portugal, which have pledged between 0.01% and 0.09% of Germany's commitments, fulfil their promises. One notable result is the fact that among countries that financed more projects there are Japan, Germany and United States and only Japan has been able to fulfil its commitments made.

3.4.2 Geographical Distribution of Climate Funds

About the 20% of total pledges in 2010, destined for biosphere protection and RES power generation, has disbursed and targeted all regions globally. Europe has received about 35% of total commitments followed by South & Central Asia and South of Sahara that received about 22% and 16% respectively. Thus, the bilateral committed aid (see Fig. 3.4) are concentrated in Europe. Focusing on disbursed funds we observe that those destined to Europe decrease and represent about the 28% of total disbursement (only Turkey received the largest allocation, around 25% of total disbursements). South & Central Asia rise up at the first place with the 33% of total disbursements shared between the 19% received by India and 7% by Nepal. The increased funds destined to this region depends on circumstance that many Asian countries are major GHG emitters and this region remains among the world's most populous.

To explore the directions and intensities of flows of funds, the analysis will concentrate only on effective disbursement made in 2010, does not considering the relate projects started in previous years. Considering that, the developing countries may receive aid from one or more donors and that, at the same time, each donor may provide financial support for several projects in one or more recipient countries, it was useful to draw a map to provide a graphic representation of the flows' directions and their corresponding amounts.

The flows reported in Figs. 3.5 and 3.6 (with a focus on European countries) show the aid directed to each country, excluding the amounts loaned by

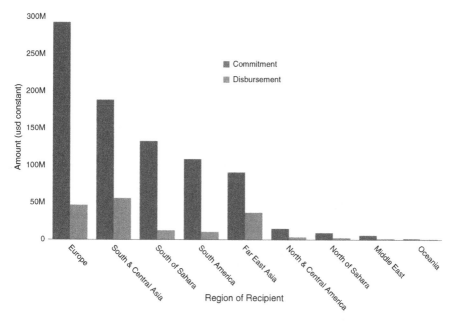

Fig. 3.4 Geographic distribution of commitment and disbursement
Source: Our elaboration on AidData Database (AidData.org)

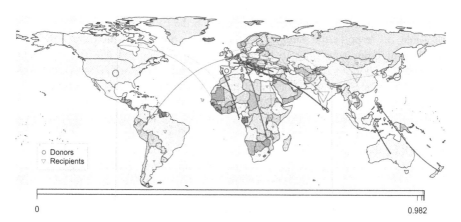

Fig. 3.5 Flow of climate finance for the year 2010. The *red* circles indicate the donors and the radius are proportional to the share of funds disbursed expressed as ratio on total climate aids; the *green* circles, represent the recipient developing countries and the radius are proportional to the share of fund received on total disbursement. The *gradient* of colored countries from straw *yellow* to *green* indicates EPI Index (from lower to cleaner countries)
Source: Our elaboration on AidData Database (AidData.org)

Fig. 3.6 Flow of climate finance for the year 2010 with focus on European countries. The *red* circles indicate the donors and the radius are proportional to the share of funds disbursed expressed as ratio on total climate aids; the *green* circles, represent the recipient developing countries and the radius are proportional to the share of fund received on total disbursement. The *gradient* of colored countries from straw *yellow* to *green* indicates EPI Index (from lower to cleaner countries)
Source: Our elaboration on AidData Database (AidData.org)

international programs and/or destined to regions or areas of the world. In the same figure, one can note the distribution of countries that are not included in Annex-II (that is, treated and untreated) and their level on the EPI.

Table 3.1 reports the ranking of the donor countries, which are sorted in descending order according to the total amount of donated aid and the main recipient to whom the funds are directed. The column "Donor/GT" represents the share of resources of the donor out of the total amount of funds disbursed by all developed countries that joined Climate Finance, while "Recipient/GT" shows the share of the recipient's received resources out of the total amount of funds disbursed. Similarly, the column "Donor/GT_donor" shows the share of resources financed by the donor towards its main beneficiary out of the total amount of funds disbursed by the donor; similarly, for the column "Recipient/GT_recipient," we calculated the share of allocated funds from the donor out of the total received aid.

The data show, both in Table 3.1 and in Figs. 3.6 and 3.7, the total amount of international funds. These funds come from the European countries and their numbers of flows are much smaller than the number of flows come from the United States. Furthermore, the main flows are directed towards developing countries that are major emerging national economies characterized by fast-growing economies. The main beneficiary of funds is Turkey, financed especially by European countries. France is the world's largest provider of climate funds, allocating 25.4% of total disbursements (approximately 165 million dollars) and guiding about 99% of its total resources (equivalent to approximately 42 million dollars) towards Turkey. All this falls under the Agence française de dévelopement's (AFD 2015) action in Turkey.

Table 3.1 Ranking of the donor countries that are sorted in descending order according to the total amount of donated aids and the main recipient to whom the funds are destined. The column "Donor/GT" represents the share of resources of donor on the total amount of funds disbursed by all developed countries that joined to Climate Finance while "Recipient/GT" shows the share of recipient's received resources on the total amount of funds disbursed. Similarly, the column "Donor/GT_donor" shows the share of financed resources by donor towards its main beneficiary on the total amount of funds disbursed by donor, similarly for the column "Recipient/GT_recipient" it was calculated the share of allocated funds from donor in the total received aids

Donor	Total of Donor[a]	Donor/GT	Donor/GT of Donor	n° financed projects by Donor	Main recipient	Recipient/GT	Recipient/GT of recipient	n° financed project in recipient
France	$ 42.1	25.45%	98.82%	3	Turkey	25.21%	99.75%	4
Germany	$ 40.4	24.45%	58.14%	84	India	18.89%	75.26%	17
Japan	$ 19.3	11.67%	37.28%	107	Nepal	7.02%	62.03%	10
United Kingdom	$ 16.1	9.76%	40.16%	24	China	13.27%	29.54%	59
United States	$ 14.3	8.63%	15.98%	45	China	13.27%	10.39%	59
Norway	$ 9.7	5.84%	35.45%	39	Nepal	7.02%	29.50%	10
Spain	$ 6.2	3.78%	11.05%	47	Cape Verde	0.97%	42.86%	3
Belgium	$ 6.2	3.74%	60.96%	9	Peru	2.61%	87.34%	7
Switzerland	$ 4.9	2.95%	32.15%	20	Bangladesh	1.09%	86.72%	3
Austria	$ 2.7	1.63%	43.92%	12	Ukraine	0.89%	80.12%	6
Netherlands	$ 2.3	1.41%	84.41%	2	Zambia	1.29%	92.21%	3
Canada	$ 0.7	0.40%	43.14%	8	Senegal	0.38%	45.90%	5
Finland	$ 0.5	0.31%	33.28%	6	United Republic of Tanzania	0.68%	15.30%	10
New Zealand	$ 0.5	0.30%	86.02%	4	Philippines	1.34%	19.17%	14
Australia	$ 0.3	0.16%	97.36%	2	Indonesia	2.42%	6.45%	15
Italy	$ 0.2	0.13%	100.00%	1	Lebanon	0.16%	81.92%	3
Portugal	$ 0.1	0.04%	100.00%	1	Sao Tome and Principe	0.04%	100.00%	1
Republic of Korea	$ 0.1	0.04%	17.97%	8	Lao People's Democratic Republic	0.05%	13.36%	7
Greece	$ 0.1	0.04%	100.00%	1	Sri Lanka	0.07%	49.19%	6

[a] Amount of funds is expressed in millions of Dollars

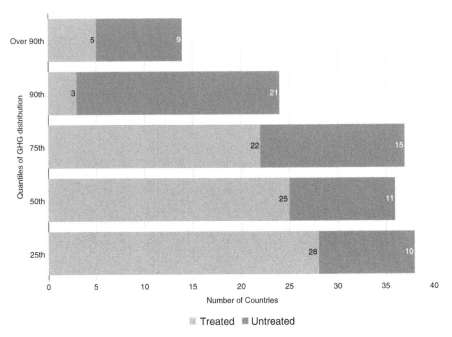

Fig. 3.7 Number of treated and untreated countries on the basis of quartiles of GHG emission distribution
Source: Our elaboration on AidData Database (AidData.org)

Box 3.9 Agence française de dévelopement (AFD) projects in Turkey
The Agence française de dévelopement's (AFD 2015) action in Turkey strenghts the French-Turkish partnership on climate and environmental issues. The aim is to provide joint responses to preserve Mediterranean ecosystems and limit greenhouse gas emissions. Between 2005 and 2015, AFD Group has allocated some EUR 2bn of financing in the country. The cooperation between France and Turkey is based on a long-standing relationship, close consultation other major issues of international and regional agendas, and significant levels of trade. Three are the important projects:
   (i) improve the sustainable transportation to support the growth of Istanbul; *(ii)* preserve the forest in Turkey and *(iii)* reduce fossil energy consumption of Turkish growth.

The second largest donor is Germany, which allocates about 24.4% of the total funding made available by all developed countries. Over half of all financial resources (58.14%) allocated by Germany for the promotion of environmental and energy projects are directed to India. Despite the country's rapidly growing economy, poverty and social issues remain a challenge and the deutsche Gesellschaft Für Internationale Zusammenarbeit (German Society for International

> **Box 3.10   Gesellschaft für Internationale Zusammenarbeit (GIZ) projects in India**
> The Deutsche Gesellschaft für Internationale Zusammenarbeit (GIZ) GmbH works jointly with partners in India for sustainable economic, ecological, and social development, focusing on energy, environment, preservation, and sustainable use of natural resources and sustainable urban development. The initiatives of GIZ support local needs and achieve sustainable and inclusive development by projects such as Smart Cities, Clean India and Skill India.

> **Box 3.11   Japan International Cooperation Agency (JICA) projects in Nepal**
> The Japan International Cooperation Agency (JICA) is advancing its activities around the pillars of a field-oriented approach, human security, and enhanced effectiveness, efficiency, and speed. The JICA Partnership Program (JPP) was introduced in 2002 to support and cooperate with the implementation of projects formulated by "Partners in Japan". In Nepal, 11 JPP projects have been implemented since the beginning of JPP. The target issues are nutrition for mothers and children, quality development of public schools, and peace education for the youth (i.e., Disaster Preparedness and Sustainable Livelihood Development Project in Chitwan District (Shapla Neer/ RRN) or Technical Cooperation in Dissemination of Alternative Energy Fuel for Firewood and Kerosene (Hokkaido Higashikawa Town/NEPA/ CEEN)).

Cooperation) is contributing to several initiatives to address India's environmental and social changes.

Japan is third in the ranking of the top donors, with a budget of 19.3 million dollars (approximately 11.7% of the total amount). It principally finances Nepal.

In addition, it is useful to note that there are also great differences in the numbers of financed projects. For example, the USA, with its 45 financed projects, does not account for even 9% of the total disbursement while France, with only three plans, reached 26%. Japan, with 107 financed projects, is in third place in this ranking. China, which received the maximum number of financed projects, receives only 13.3% of total funding. Turkey has received approximately 25% of total funding with only 4 "green-growth" financed projects.

3.4.3   Disbursement and GHG Emissions

As assessed by UNFCCC (2014) the financial resources directed to developing countries for mitigation and adaptation activities will become increasingly significant. More and more countries have begun to take on more responsibility for

Box 3.12 Environmental Pollution Index (EPI)
The most important anthropogenic emissions of greenhouse gases (GHG): Carbon Dioxide (CO2), Methane (CH4), Nitrous oxide (N2O) and Chloro-fluorocarbons (CFCs), hydro chlorofluorocarbons (HCFCs), hydrofluorocarbons (HFCs), perfluorocarbons (PFCs), and sulfur hexafluoride (SF6), together called F-gases, are the indicators that will be combined in a composite indicator.

The indicators are measured in the same units and have the same direction. The data are scaled, using the min-max method, and is accorded to the high and low values. We obtain an Environmental Pollution Index (EPI) that puts the "cleaner" countries at the higher levels of the index and the major polluters at the lower levels.

To obtain the EPI, we aggregate the weights for each indicator using the geometric methods that reduce the compensability of indicators (OECD 2008):

$$CI_c = \prod_{q=1}^{Q} X_{q,c}^{w_q}$$

For details on its construction, see the chapter of Methods.

mitigating GHG emissions to combat the impacts of climate change resulting from their economic development.

Thus, the analysis focus on the flows of funds among countries and on the relationship between the "Fast-Start finance" and GHG emission (summarize with the Environmental Pollution Index (EPI), see Box 3.12 for details) to assess whether funds primarily reach the main polluting developing countries and to determine whether any preferences could undermine the effectiveness of climate finance.

The Fig. 3.7 shows the distribution of funding among countries having regard to the quantile of GHG emission distribution. The bar chart reveals that 28 of 149 analyzed countries have relatively high GHG emissions but there are, also, ten countries that are eligible and not received funding.

While there are five countries which have very low GHG emissions and received them. Observing the Fig. 3.7 we can note that within the 25th quantile of distribution of GHG emission there are countries that have received no allocations and, among those have received don't match with amount of GHG emissions. However, observing the Fig. 3.8, the major allocations are directed towards the countries with higher level of GHG emission.

A comparison of the geographic distribution of bilateral climate finance with the GHG emissions of recipient countries (see Fig. 3.9) using data from the Climate Analysis Indicator Tool (CAIT, using data for the most recently reported year: 2012) suggests that funding are more concentrated in countries with higher

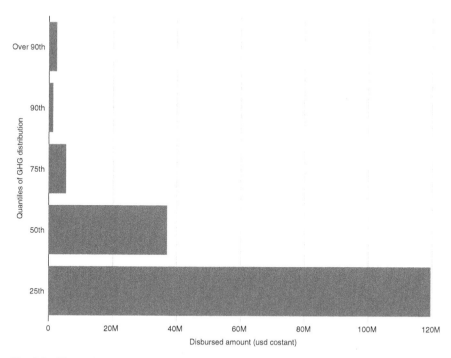

Fig. 3.8 Climate funds received by developing countries
Source: Our elaboration on AidData Database (AidData.org)

emissions. In Nakhooda and Norman (2014), also climate funding (total amount between 2003 and September 2014) is heavily concentrated in countries with high and growing emissions and on the top 3 recipients of funding we found three of the largest GHG emitters in the world: India, Brazil and Indonesia. Thus, with regarding the analyzed aid, there are countries like China, Indonesia and Brazil that have very high emission, but received lower than countries as Turkey that have lower emission.

In our analysis, (only bilateral funds disbursed in 2010 and only on two type of aid) on the top 3 recipients there are Turkey, India and Nepal and they aren't among the largest GHG emitters in the world (excluding India). Another notable difference is in type of funding received: Turkey received only aid targeted to biosphere protection, while in India, a fast-growing economy, the funding are destined to develop energy production.

When we compare the geographic distribution of "Fast-start finance" and a summary of development of countries (we have considered as proxy the GDP and the Electricity consumption) we observe (see Fig. 3.10) that wealthy developing countries also have received, thus reflecting the observation made above.

As shown in Figs. 3.6, 3.7, 3.9 and 3.10 the financial flows from "Fast-start Finance" mainly go to rapidly developing countries. As suggested by Pickering et al. (2015) and Markandya et al. (2015), the main issue concerns the need to

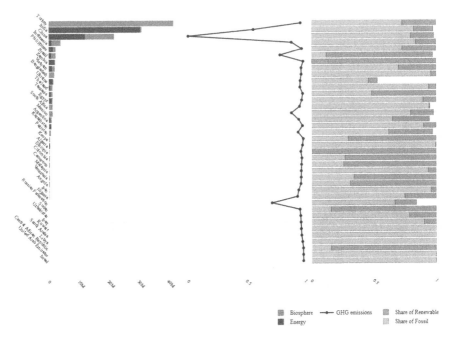

Fig. 3.9 Disbursement and environmental performance of recipients
Source: Our elaboration on AidData Database (AidData.org)

enhance the international negotiations process and inter-agency dynamics to enable funds to be directed towards specific needs and priorities. Furthermore, the donors give greater weight to political influence against climate deterioration partly because they pursue international strategies but also because they have the required influence to be able to assert their power (see, e.g., Bigsten and Tengstam 2015).

Moreover, the heterogeneity in the number and amount of financed projects makes it difficult to fully evaluate and understand the effectiveness of individual aid awards and programs or to distinguish between the different types of projects and choose the best way of directing financial aid to promote "green growth" more effectively.

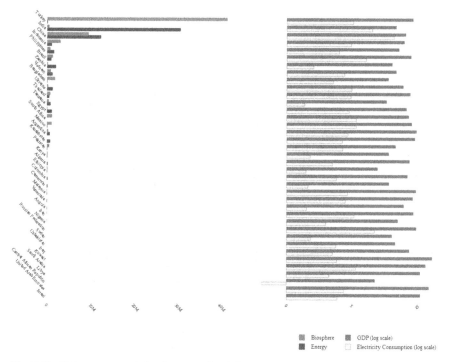

Fig. 3.10 Disbursement and development of recipients
Source: Our elaboration on AidData Database (AidData.org)

References

Agence Française de Développement, "AFD's Strategy in Turkey", (2015) Available at http://www.afd.fr/webdav/site/afd/shared/PUBLICATIONS/THEMATIQUES/AFD_Turquie_GB.pdf. Accessed 07 Oct 2016

Bazilian M, Nussbaumer P, Gualberti G, Haites E, Levi M, Siegel J, Fenhann J (2011) Informing the financing of universal energy access: An assessment of current financial flows. Electr J 24 (7):57–82. doi:10.1016/j.tej.2011.07.006

Bhattacharyya SC (2013) Financing energy access and off-grid electrification: A review of status, options and challenges. Renew Sust Energ Rev 20:462–472. doi:10.1016/j.rser.2012.12.008

Bigsten A, Tengstam S (2015) International coordination and the effectiveness of aid. World Dev 69:75–85. doi:10.1016/j.worlddev.2013.12.021

Bird N, Tilly H, Trujillo NC, Tumushabe G, Welham B, and Yanda P (2013) Measuring the effectiveness of public climate finance delivery at the national level, ODI; London. http://www.odi.org.uk/sites/odi.org.uk/files/odi-assets/publications-opinion-files/8303.pdf. Accessed 08 Oct 2016]

Bourguignon F, Platteau JP (2015) The hard challenge of aid coordination. World Dev 69:86–97. doi:10.1016/j.worlddev.2013.12.011

Brunner S, Enting K (2014) Climate finance: a transaction cost perspective on the structure of state-to-state transfers. Glob Environ Chang 27:138–143. doi:10.1016/j.gloenvcha.2014.05.005

Ellis J, Caruso R, Ockenden S (2013) Exploring climate finance effectiveness, Climate Change Expert Group Paper No. 2013(4). OECD, Paris

Espagne E (2016) Climate finance at COP21 and after: lessons learnt, CEPII, Policy Brief, No 9 – February 2016. CEPII research center, Paris. http://www.cepii.fr/PDF_PUB/pb/2016/pb2016-09.pdf. Accessed 17 Oct 2016

Indipendent Commission for Aid Impact (2014) The UK's international climate funds. ICAI Report 38 http://icai.independent.gov.uk/wp-content/uploads/ICAI-Report-International-Climate-Fund.pdf. Accessed 14 Feb 2017

Keeley AR (2016) Renewable energy in pacific small island developing states: the role of international aid and the enabling environment from donor's perspectives. J Clean Prod. Available online 11 May 2016, doi: 10.1016/j.jclepro.2016.05.011. Accessed 3 Oct 2016

Markandya A, Antimiani A, Costantini V, Martini C, Palma A, Tommasino MC (2015) Analyzing trade-offs in international climate policy options: The case of the green climate fund. World Dev 74:93–107

Marquardt J, Steinbacher K, Schreurs M (2016) Driving force or forced transition?: The role of development cooperation in promoting energy transitions in the Philippines and Morocco. J Clean Prod 128:22–33. doi:10.1016/j.jclepro.2015.06.080

Nakhooda S (2013) The effectiveness of climate finance: a review of global environment facility. Overseas Development Institute, London

Nakhooda S, Norman M (2014) Climate finance: is it making a difference? a review of the effectiveness of multilateral climate funds. Overseas Development Institute, London

Nakhooda S, Fransen T, Kuramochi T, Caravani A, Prizzon A, Shimizu N, Tilley H, Halimanjaya A, Welham B (2013) Mobilising international climate finance: lessons from the fast-start finance period. Overseas Development Institute, London

OECD (2008) Handbok on constructing composite indicators. Methodology and User Guide. OECD, Paris

OECD (2015) Climate finance in 2013–14 and the USD 100 billion goal. OECD, Paris

Pickering J, Skovgaard J, Kim S, Roberts JT, Rossati D, Stadelmann M, Reich H (2015) Acting on climate finance pledges: Inter-agency dynamics and relationships with aid in contributor states. World Dev 68:149–162

Tierney MJ, Nielson DL, Hawkins DG, Timmons Roberts J, Findley MG, Powers RM, Parks B, Wilson SE, Hicks RL (2011) More dollars than sense: refining our knowledge of development finance using AidData. World Dev 39(11):1891–1906

Tirpak D, Brown L, Ronquillo-Ballesteros A (2014) Monitoring climate finance in developing countries: challenges and next steps. World Resources Institute, Washington DC

Tol RS (2017) The structure of the climate debate. Energy Policy 104(May 2017):431–438. ISSN 0301-4215, http://dx.doi.org/10.1016/j.enpol.2017.01.005

UNFCCC (2014) Biennial assessment and overview of climate finance flows 2014. Accessible at: http://unfccc.int/cooperation_and_support/financial_mechanism/standing_committee/items/2807.php

UNFCCC (2016b) Report of the conference of the parties on its twenty-first session, held in Paris from 30 November to 13 December 2015. Addendum. Part two: Action taken by the Conference of the Parties at its twenty-first session. United Nations Office at Geneva, Geneva

UNFCCC (2016a) Summary and recommendations by the standing committee on finance on the 2016 biennial assessment and overview of climate finance flows. Biennial Assessment and Overview of Climate Finance Flows

Urpelainen J (2012) Strategic problems in North–South climate finance: creating joint gains for donors and recipients. Environ Sci Pol 21:14–23. doi:10.1016/j.envsci.2012.03.001

Zhao SX, Chan RC, Chan NYM (2012) Spatial polarization and dynamic pathways of foreign direct investment in China 1990–2009. Geoforum 43(4):836–850. doi:10.1016/j.geoforum.2012.02.001

Chapter 4
Assessing the Effectiveness of Climate Finance: Composite Indicators and Quantile Regression

Abstract In this chapter we briefly explain methods proposed to assess the effectiveness of the climate funds to reduce the greenhouse gas emissions.

Greenhouse gas emissions are constituted by Carbon Dioxide (CO_2), Methane (CH_4), Nitrous oxide (N_2O) and Chlorofluorocarbons (CFCs), hydro chlorofluorocarbons (HCFCs), hydrofluorocarbons (HFCs), perfluorocarbons (PFCs), and sulfur hexafluoride (SF_6), together called F-gases.

Considering the multidimensional concept and the complexity of the topic we first explain the composite indicator as useful tools to summarize in one indicator the individual components of the greenhouse gas emissions and to compare performances of the countries. Second, we describe the quantile regression, focusing on the theory and possible applications in research fields of interest. Quantile regression, in fact, is one of the tools used to assess the effects of economic policies on some of the target variables.

Keywords Greenhouse gas • Composite indicators • Principal components analysis • Weighting methods • Quantile regression • Clustered data

4.1 Introduction

Greenhouse gas emissions (GHG) are gases that trap heat in the atmosphere and their effect on climate change depends on three main factors:

How much of these gases are in the atmosphere?
How long do they stay in the atmosphere?
How strongly do they affect global temperatures?
How we measure GHG emissions can importantly influence how we come to understand it, how we analyze it, and how we propose policies to reduce it.

Literature generally proposes the use of carbon dioxide (CO_2) emission as a proxy that captures the environmental degradation due to economic development to test the effect of GHG on some outcome (see, e.g., van Ruijven and van Vuuren 2009; Marques and Fuinhas 2012) or, alternatively, to explain how covariates can influence it (see, e.g., Liu 2005; Mundaca et al. 2015).

Nevertheless CO_2 is the primary GHG emitted through anthropogenic activities and account for about 65% of global GHG emissions (IPCC 2014) and represents about three quarters of the anthropogenic GHG emissions for Annex I countries (IEA 2013) it is certainly not the only cause of climate change. For this reason, we consider the most important anthropogenic emissions of greenhouse gases (GHG): Carbon Dioxide (CO_2), Methane (CH_4), Nitrous oxide (N_2O) and Chlorofluorocarbons (CFCs), hydro chlorofluorocarbons (HCFCs), hydrofluorocarbons (HFCs), perfluorocarbons (PFCs), and sulfur hexafluoride (SF6), together called F-gases.

Considering the multidimensional concept and the complexity of the topic we first propose a composite indicator to assess the impact of anthropogenic activities and to rank the countries and, second, we study the relationship between the proposed indicator and funds provided by donor countries to promote green growth in developing countries using a quantile regression model.

In this chapter, we briefly explain methods employed to summarize in one indicator the individual components of the GHG and to study the relationship between this indicator and the climate funds.

4.2   Composite Indicator: An Introduction

Composite indicators (CIs) are considered useful tools to compare performances. Especially in the area related to the geographical (e.g., country), they serve as instruments in policy analysis to communicate and compare complex issues. They can also be helpful in setting policy priorities and in benchmarking or monitoring performance. A composite indicator is formed when individual indicators are compiled into a single index based on an underlying model. The composite indicator should ideally measure multidimensional concepts, which cannot be captured by a single indicator, e.g. competitiveness, industrialization, sustainability, single market integration, knowledge-based society, etc.

Aim of this section is to summarize the approaches commonly employed for constructing a composite indicator. Major details can be found in the *Handbook on Constructing Composite Indicators* by OECD (2008). Furthermore, the *COMPIND* package of the *R* statistical software contains functions to enhance several approach to the composite indicator methods, focusing on normalization and weighting – aggregation steps.

Summarizing and according to OECD (2008), "A composite indicator (CI) is formed when individual indicators are compiled into a single index, on the basis of an underlying model of the multi-dimensional concept that is being measured".

One of the virtue of composite indicators is their usefulness to summarize complex and sometimes elusive issues in wide ranging fields and that they often seem easier to interpret than finding a common trend in many separate indicators and have proven useful in benchmarking country performance. However, composite indicators builders have to face a relevant degree of skepticism among statisticians, economists and other groups of users. This skepticism is partially due to the lack of transparency of some existing indicators, especially as far as methodologies and basic data are concerned. Composites can send misleading policy messages if they are poorly constructed or misinterpreted. Their "big picture" results may invite users (especially policy makers) to draw simplistic conclusions. On the other hand, a sound process of construction cannot remedy to an inadequate framework or to poor quality data. Thus, the creation of a composite indicator requires a balance between different aspects, all equally important in defining the quality and finally the usefulness of the composite. The main pros and cons of using composite indicators can be found in Saisana and Tarantola (2002) or in OECD (2008).

To better design, develop and disseminate a CI it needs to follow a set of recommendations. In particular, Handbook on constructing composite indicators by OECD (2008) to guide constructors discusses the following steps in the construction of composite indicators:

- Theoretical framework;
- Data selection;
- Imputation of missing data;
- Multivariate analysis;
- Normalization;
- Weighting and aggregation.

A theoretical framework should be developed to provide the basis for the selection and combination of single indicators into a meaningful composite indicator under a fitness-for-purpose principle. Indicators should be selected on the basis of their analytical soundness, measurability, accessibility, country coverage, relevance to the phenomenon being measured and relationship to each other. The use of proxy variables should be considered when data are scarce.

One of the main problem is given by the missing data. Missing data are often a problem in in considering international data. Dealing with nonresponse can be a difficult matter and it is important to apply adequate missing data methods to obtain valid inference.

Different are the approaches generally employed for imputing missing values. They can be summarize in: (i) case deletion; (ii) single imputation; (iii) multiple imputation. Following a general rule of thumb suggested by Little and Rubin (2002), cases are not deleted if a variable has more than 5% of missing values. Simple imputation methods are commonly used but these may not be adequate in many circumstances (see, e.g., Ibrahim et al. 2005). More sophisticated methods, such as fractional hot deck, multiple imputation and generally less parametric methods, have advantages and the use of such advanced methods may be preferable. Overviews of such methods are given in Allison (2001), GSS (1996), Quintano and Castellano (2001), Little and Rubin (2002), Schafer and Graham (2002), Ibrahim et al. (2005).

A preliminary multivariate analysis of the data has to be carefully conducted before the construction of a CI. This step helps to assess the suitability of the data set and explain the methodological choices for e.g. weighting and aggregating the single indicator (OECD 2008) suggests that information can be grouped and analyzed along at least two dimensions of the data set: individual indicators and countries.

Successively, the selected indicators should be normalized prior to the data aggregation to render them comparable. Practitioners should take into account data properties, as well as the objectives of the CI in choosing one of the various normalization method (see, e.g., Freudenberg 2003; Jacobs et al. 2004). One has to be aware that different normalization methods will yield different results in composite scores. Normalization techniques are summarized and discussed in Nardo et al. (2004).

Attention needs to be paid to extreme values as they may influence subsequent steps in the process of building a CI.

The focal point in the construction process of a CI is represented by the need to combine different indicators. This implies the choice of the weighting method will be applied to aggregate the information according to the underlying theoretical framework. Correlation and compensability issues among indicators need to considered and either be corrected for or treated as features of the phenomenon that need to retained in the analysis.

The weighting procedure requests that weights should reflect the contribution of each indicator to the overall composite and it is particularly delicate when the phenomena we want to summarize and ranking is complex, interrelated and multidimensional.

One of the most employed weighting method is based on the same weight associated to the normalized variables. This method is based on the basic assumption that all the variables considered relevant by the constructors have the same importance in the CI or, alternatively, it can be the results of insufficient knowledge of causal relationships (like in the case of 2002 Environmental Sustainability Index). In other case, the combination of variables with a high degree of correlation can suggest to introduce some other methods to weigh the indicators based on statistical models such as principal components analysis or factor analysis (see, e.g., Nicoletti et al. 2000).

In this case, the statistical methods account for the highest variation in the data set, using the smallest possible number of factors that reflect the underlying "statistical" dimension of the data set. Weighting only intervenes to correct for the overlapping information of two or more correlated indicators, and it is not a measure of the theoretical importance of the indicators. Weights, however, cannot be estimated if no correlation exists between indicators. Other statistical methods that employ linear programming tools, such as the benefit of the doubt (BOD) are extremely parsimonious about weighting assumptions as it lets the data decide on the weights and is sensitive to national priorities (Melyn and Moesen 1991; Cherchye et al. 2008). However, given that the weights are derived from the data, they are also subject to eventual data measurement errors.

Other weighting procedure could be based on participatory methods that incorporate various stakeholders to assign weights to single variables. As suggested by

Munda (2005a, b, 2007) the participatory approach is feasible if there is a well-defined basis for a national policy.

Other weighting methods are proposed by the literature such as:

- Weighting method based on geometric aggregation
- Mazziotta-Pareto Index (MPI) method
- Wroclaw Taxonomic Method

OECD (2008) suggests the good practice to use different weighting methods and test their effect on final country scores (and ranking). Weights usually have an important impact on the value of the composite and on the resulting ranking especially whenever higher weight is assigned to individual indicators on which some countries excel. Moreover, assumptions ad implication of weighting methods employed to measure the importance of each variables should be always made clear and tested for robustness.

Weighting methods are strongly related to how the variables are aggregated into a CI. Numerous aggregation methods are possible and they can be based on linear or nonlinear algorithmic. Each technique implies different assumptions and has specific consequences.

When all variables, or sub-indicators, have the same measurement unit and the ambiguities on scale effects is eliminated then is useful to employ a linear aggregation method. Generally, it is assumed that linear aggregation.

Geometric aggregations, instead, are appropriate when non-comparable and strictly positive individual indicators are expressed in different ratio-scales. The absence of synergy or conflict effects among the indicators is a necessary condition to admit either linear or geometric aggregations. Furthermore, linear aggregations reward base indicators proportionally to the weights, while geometric aggregations reward more those countries with higher scores.

Linear and geometric aggregation criterions assume that deficits in one dimension can be offset by the surplus in another (Munda and Nardo 2007). Compensability is assumed constant if the linear aggregation is employed while it is assumed lower when the composite contains indicators with low values with the geometric aggregation.

In some cases, however, the presence of different goals are important then a non-compensatory logic has to be followed. The non-compensatory multicriteria approach will assure non-compensability by formalizing the idea of finding a compromise between two or more legitimate goals.

Table 4.1 summarizes the relationship between aggregation and weighting methods indicating the combination of weighing and aggregation that guarantees no loss of mathematical properties.

Last step of the construction process is the assessment of robustness and sensitivity. This analysis has the aim to test how the CI respond to change in normalization criterion and/or weighting and aggregation methods employed.

Table 4.1 Compatibility between aggregation and weighting methods

Weighting methods	Aggregation methods		
	Linear[a]	Geometric[a]	Multi-criteria
EW	Yes	Yes	Yes
PCA	Yes	Yes	Yes
BOD	Yes[b]	No[c]	No[c]
UCM	Yes	No[c]	No[c]
BAP	Yes	Yes	Yes
AHP	Yes	Yes	No[d]
CA	Yes	Yes	No[d]

Source: OECD (2008)

[a]With both linear and geometric aggregations weights are trade-offs and not "importance" coefficients

[b]Normalized with the Min-Max Method

[c]BOD requires additive aggregation, similar arguments apply to UCM

[d]At least with the multi-criteria methods requiring weights as importance coefficients

4.2.1 A Composite Indicator for GHG Emissions: The Environmental Pollution Index

The construction of an index that can summarize the GHG emissions of countries is complex and uses the four variables that have been previously identified. One way to make comparison across space (and time) is to combine the various indicators in a single index.

A CI has the advantage of allowing the ranking of countries because it represents overall Environmental Performance in one number. Nevertheless, building composite indices implies losing a certain amount of information.[1] However, as monitoring GHG emissions often requires overall comparisons across space, composite indices are very useful for specific purposes.

The CO_2, N_2H, CO_4 and F-gases indicators will be combined in a composite indicator. As suggested by OECD (2008), when missing data were detected we employed the imputation method based on regression.

The indicators are measured in the same units and have the same direction. However, following the suggestions of the OECD, we have first normalized the indicators using the min-max method.[2] In this way, we ensure comparability among countries. The data are scaled according to the high and low values, which represent the possible range of a variable for all countries. We obtain an Environmental Pollution Index (EPI) that puts the "cleaner" countries at the higher levels of the

---

[1]Furthermore, composite indices have been criticized because, in a way, they re-introduce unidimensionality. For the pros and cons of using CIs see, e.g., Nardo et al. (2004) and OECD (2008).

[2]There is a wide range of normalization methods (see, e.g., OECD 2008) and the choice depends on the type of data and on weighting and aggregation (Hudrilikova 2013).

index and the major polluters at the lower levels. In this way, we evaluate the effectiveness of funds (in the following section) at improving the performances of countries. According to the direction of the elementary indicators, the following min-max formula ((4.1)) has been employed:

$$I_{qc} = \frac{max_c\left(x_q\right) - x_{qc}}{max_c\left(x_q\right) - min_c\left(x_q\right)}, \tag{4.1}$$

where x_{qc} is the value of an indicator q for country c. For joint normalization between component indicators, the scales of each component are converted linearly in such a way that the lowest score among individual countries for a component indicator is set to 0 and the highest score is set to 1. The scores vary between the theoretical lower and upper bounds of 0 and 1.[3]

Once all indicators are transformed into the 0–1 interval, the measure of each indicator of the EPI is computed by aggregating the component indicators following two so-called statistical methods:

- Equal weighting (EW)
- Principal component analysis (PCA)

Using the EW methods, each indicator is assigned the same weight, or:

$$w_q = 1/Q, \tag{4.2}$$

where w_q is the weight for the q-th indicators ($q = 1, .., Q$) for each country.

The other weighting method is based on Principal Component Analysis (PCA). PCA is a multivariate technique that is used to reduce the dimensionality of a dataset containing a number of interrelated variables[4] while retaining as much of the variation of the data as possible (Doukas et al. 2012). It allows transforming an original set of variables into a new set, including linear combinations of the original variables, with a smaller dimension but most of the variation of the original set. The application of PCA requires multiple steps.

After the normalization of the indicators, the eigenvalues, λ, and the eigenvectors, F, of the correlation matrix R of the selected indicators are computed. This has been obtained through the following determinant equation:

$$(R - \lambda I) = 0, \tag{4.3}$$

where I is the unit matrix.

---

[3]Extreme values (outliers) are identified but not adjusted for, at least not in this first version of the composite. Nor has it been considered necessary at this stage to make any non-linear transformations of any of the underlying component indicators.

[4]PCA is usually employed in explanatory data analysis frameworks as well as for developing predictive models and a wide range of applications concerned with economic, social and environmental aspects.

Solving Eq. (4.3), Q eigenvalues corresponding to the correlation matrix R are calculated. The eigenvalue with the largest rate is the one that holds most of the variation and the eigenvalues with very small rates are usually ignored. Furthermore, to derive the eigenvectors, the following equation is solved:

$$(R - \lambda_j I)F_j = 0 \tag{4.4}$$

where F_j is the matrix of the eigenvectors corresponding to the λj eigenvalues. Finally, a matrix whose columns are the calculated eigenvectors is transposed and multiplied for the original normalized variable j.

By this computation, the replacement of the set of indicators by an adequate number of Principal Components (PCs) is performed. The PCs are considered normalized linear functions of the indicator variables and are mutually orthogonal. The first component accounts for the largest proportion of total variation (trace of the covariance matrix) of all indicator variables. The second accounts for the second largest proportion, and so on.

Taking into account that an eigenvalue represents the variance of the corresponding PC, it can be employed as weight. Thus, the composite indicator is developed using the variances of the calculated PCs as weights.

To obtain the EPI, we aggregate the weights for each indicator using the geometric methods that reduce the compensability of indicators (OECD 2008):

$$CI_c = \prod_{q=1}^{Q} X_{q,c}^{w_q} \tag{4.5}$$

4.3   Quantile Regression with Cluster Data

Regression-based causal inference is predicated on the assumption that when key observed variables have been made equal across response variable and predictor variables, selection bias from the things we cannot see is also mostly eliminated. The basic regression model, the linear model, using the method of the least squares, allows estimating the conditional mean of the response variable. Another basic assumption regards the distributions of both the response variable and disturbances. When these assumptions fall the least squares can be severely distorted by outlying observations and all the parametric methods may be inconsistent. Some recent applications have led to the proposal of "robust semiparametric estimators" that have unaffected by the presence of outliers or high leverage observations in the data. Semiparametric estimators are based on fewer assumptions respect to the parametric estimators: they depend by the characteristics of the population and, in general, the distributional assumptions are removed. This is an important advantage because makes those more robust than the parametric estimator. The principal disadvantage regards the estimation methods that may be more complexes also in computationally terms. Among the semiparametric estimators, the least absolute deviation estimator (LAD) has been suggested as an alternative that remedies the presence of to heavy-tailed distributions and outliers.

The LAD estimator is the solution to the optimization problem:

$$Min_b \sum_{i=1}^{n} | y_i - x_i'b |$$

The LAD estimator is a special case of quantile regression:

$$prob\left[y_i \leq x_i'\beta\right] = q$$

The LAD estimator estimates the median regression, it is the solution of the quantile regression when $q = 0.5$.

An extension of LAD estimator is the quantile regression model that, under certain assumptions, aims at estimating either the conditional median or other quantiles of the response variable. The Quantile regression, which was originally introduced by Koenker and Bassett (1978), models the relationship between X and the conditional quantiles of Y, and it is useful in applications where extremes are important (see, e.g., Yu et al. 2003; Ranganai et al. 2014), or in case of asymmetric distribution such as environmental studies (see, e.g., Davino et al. 2014), and least squares can be severely distorted by them. The great advantage of quantile regression is to generalize the concept of a univariate quantile to a conditional quantile given one or more covariates. Furthermore, quantile regression is robust to heavy-tailed distributions and outliers. Other advantages of quantile regression are summarized in Buchinsky (1998). Koenker and Bassett (1978) propose, for the calculation of quantiles, an alternative procedure to the classical one. In particular, this procedure is based on an optimization method. More precisely, analogous to what happens to the sample mean, which can be defined as the solution of the minimization problem of the sum of squared deviations, here we can define each quantile as the solution of the following minimization problem.

For $0 \leq \alpha \leq 1$ the α-th quantile of y given x is defined by

$$Q_y(\alpha|x) = \min\{\eta|P(y \leq \eta|x) \geq \alpha\} \tag{4.6}$$

In addition, assuming that $Q_y(\alpha|x)$ is linear, we obtain that

$$Q_y(\alpha|x) = x'\beta(\alpha)$$

which is equivalent to

$$y = x'\beta(\alpha) + u(\alpha) \tag{4.7}$$

$$Q_y(\alpha|x) = x'\beta(\alpha) \tag{4.8}$$

Parente and Santos Silva (2016) extend the results of Kim and White (2003) and show that the traditional quantile regression analysis is consistent and asymptotically normal when there is intra-cluster correlation of the error terms. Based on this finding, we implement a quantile regression with robust standard errors, which allows controlling for the case in which the error may be clustered (see Parente and Santos Silva 2016).

In other words, let the data be $\{(y_{gi}, x_{gi}), g = 1, \ldots, G; i = 1, \ldots, n_g\}$, where g indexes a set of G clusters, each with n_g elements. We assume that the disturbances are conditionally independent across clusters, but can be correlated within clusters (Parente and Santos Silva 2016). Therefore, the model to be estimated is:

$$y_{gi} = x'_{gi}\beta(\alpha) + u(\alpha)_{gi} \tag{4.9}$$

and $\beta(\alpha)$ can be estimated as

$$\widehat{\beta} = \arg\min_b \frac{1}{G} \sum_{g=1}^{G} \left\{ \sum_{y_{gi} \geq x'_{gi}b} \alpha |y_{gi} - x'_{gi}b| + \sum_{y_{gi} < x'_{gi}b} (1-b)|y_{gi} - x'_{gi}b| \right\} \tag{4.10}$$

$\widehat{\beta}(\alpha)$ is usually estimated by linear programming methods.

In a recent paper, Parente and Santos Silva (2016) extend the results of Kim and White (2003) and show that the quantile regression estimator (Koenker and Bassett 1978) is consistent and asymptotically normal when there is within-cluster correlation of the error terms. They also provide a consistent estimator of the covariance matrix and propose a specification test able to detect the presence of intra-cluster correlation. In particular, in the absence of intra-cluster correlation, the covariance estimator proposed by Parente and Santos Silva (2016) is equivalent to a standard heteroscedasticity robust estimator; this is the case when $n_g \equiv 1$ (Powell 1984; Chamberlain 1994; Kim and White 2003; Parente and Santos Silva 2016).

In this work, we use the analytical World Bank classification of the country's economies based on estimates of gross national income (GNI) per capita to define the group that may cause the intra-cluster correlation. According with the World Bank income classification, four clusters have been identified: low-income economies that are defined as those with a GNI per capita of $1025 or less. Lower middle-income economies that are those with a GNI per capita between $1026 and $4035. Upper middle-income economies that are those with a GNI per capita between $4036 and $12,475 and the high-income economies with a GNI per capita of $12,476 or more. The updated GNI per capita estimates are also used as input to the World Bank's operational guidelines that determines lending eligibility.

References

Allison PD (2001) Missing data, sage university papers series on quantitative applications in the social sciences, series no. 07–136, Thousand Oaks

Buchinsky M (1998) Recent advances in quantile regression models: a practical guideline for empirical research. J Hum Resour 33(1):88–126. doi:10.2307/146316

Chamberlain G (1994) Quantile regression, censoring, and the structure of wages. In: CA Sims (Ed.) Advances in Econometrics: Volume 1: Sixth World Congress, chapter 5, Cambridge University Press

Cherchye L, Moesen W, Rogge N, van Puyenbroeck T, Saisana M, Saltelli A, Liska R, Tarantola S (2008) Creating composite indicators with DEA and robustness analysis: the case of the technology achievement index. J Oper Res Soc 59:239–251

Davino C, Furno M, Vistocco D (2014) Quantile regression. Theory and applications. Wiley Series in Probability and Statistics. John Wiley & Sons, Ltd.

Doukas H, Papadopoulou A, Savvakis N, Tsoutsos T, Psarras J (2012) Assessing energy sustainability of rural communities using Principal Component Analysis. Renew Sustain Energy Rev 16(4):1949–1957. doi:10.1016/j.rser.2012.01.018

Freudenberg M (2003) Composite indicators of country performance: a critical assessment. OECD, Paris

Government Statistical Service (GSS) (1996) Report of the task force on imputation, government statistical service methodology series, 3, London

Hudrilikova L (2013) Composite indicators as a useful tool for international comparison: the Europe 2020 example. Prague Economic Papers 4:459–473

Ibrahim JG, Chen MH, Lipsitz SR, Herring AH (2005) Missing-data methods for generalised linear models: a comparative review. J Am Stat Assoc 100(469):332–346

IEA, World Energy Outlook 2013 (2013)

IPCC (2014) Climate change 2014: Mitigation of climate change. Contribution of working group III to the fifth assessment report of the intergovernmental panel on climate change. In: Edenhofer O, Pichs-Madruga R, SokonaY, Farahani E, Kadner S, Seyboth K, Adler A, Baum I, Brunner S, Eickemeier P, Kriemann B, Savolainen J, Schlömer S, von Stechow C, Zwickel T and Minx JC (eds). Cambridge University Press, Cambridge/New York

Jacobs R, Smith P, Goddard M (2004) Measuring performance: an examination of composite performance indicators, Centre for Health Economics, Technical Paper Series 29

Kim TH, White H (2003) Estimation, inference, and specification testing for possibly misspecified quantile regression. Adv Econ 17:107–132. doi:10.1016/S0731-9053(03)17005-3

Koenker R, Bassett GS Jr (1978) Regression Quantiles. Econometrica 46:33–50

Little RJA, Rubin DB (2002) Statistical analysis with missing data, wiley interscience. J. Wiley &Sons, Hoboken

Liu X (2005) Explaining the relationship between CO2 emissions and national income—The role of energy consumption. Econ Lett 87(3), June 2005, 325–328 http://dx.doi.org/10.1016/j.econlet.2004.09.015

Marques AC, Fuinhas JA (2012) Are public policies towards renewables successful? Evidence from European countries. Renew Energy 44:109–118

Melyn W, Moesen WW (1991) Towards a synthetic indicator of macroeconomic performance: unequal weighting when limited information is available, Public Economic research Paper 17, CES, KU Leuven

Munda G (2005a) Multi-criteria decision analysis and sustainable development. In: Figueira J, Greco S, Ehrgott M (eds) Multiple-criteria decision analysis. State of the art surveys. Springer, New York

Munda G (2005b) "Measuring sustainability": a multi-criterion framework. Environ Dev Sustain 7 (1):117–134

Munda G (2007) Social multi-criteria evaluation. Springer-Verlag, Heidelberg/New York. Economics Series

Munda G, Nardo M (2007) Non-compensatory/non-linear composite indicators for ranking countries: a defensible setting. Appl Econ 41:1513–1523

Mundaca L, Román R, Cansino JM (2015) Towards a green energy economy? A macroeconomic-climate evaluation of Sweden's CO2 emissions. Appl Energy 148, 15 June 2015, 196–209 http://dx.doi.org/10.1016/j.apenergy.2015.03.029

Nardo M, Tarantola S, Saltelli A, Andropoulos C, Buescher R, Karageorgos G, Latvala A, Noel F (2004) The e-business readiness composite indicator for 2003: a pilot study, EUR 21294

Nicoletti G, Scarpetta S, Boylaud O (2000) Summary indicators of product market regulation with an extension to employment protection legislation, OECD, Economics department working papers No. 226, ECO/WKP(99)18. http://www.oecd.org/eco/eco

OECD (2008) Handbok on constructing composite indicators. Methodology and user guide. OECD, Paris

Parente PM, Santos Silva J (2016) Quantile regression with clustered data. J Econ Methods 5 (1):1–15. doi:10.1515/jem-2014-0011

Powell JL (1984) Least absolute deviations estimation for the censored regression model. J Econ 25(3):303–325. doi:10.1016/0304-4076(84)90004-6

Quintano C, Castellano R (2001) Strategies for dealing with non-responses for quality in some ISTAT surveys, Essays-ISTAT, n 11: 1–110

Ranganai E, Van Vuuren JO, De Wet T (2014) Multiple case high leverage diagnosis in regression quantiles. Commun Stat Theory Methods 43(16):3343–3370

Saisana M, Tarantola S (2002) State-of-the-art report on current methodologies and practices for composite indicator development, EUR 20408 EN. Italy, European Commission-JRC

Schafer JL, Graham JW (2002) Missing data: our view of the state of the art. Psychol Methods 7 (2):147–177

Van Ruijven B, van Vuuren DP (2009) Oil and natural gas prices and greenhouse gas emission mitigation. Energy Policy 37(11):4797–4808

Yu K, Lu Z, Stander J (2003) Quantile regression: applications and current research areas. The Statistician 52(3):331–350

Chapter 5
Empirical Study of Climate Finance

Abstract The chapter proposes one of the possible approaches to investigate whether developing countries can progress towards more environmentally sustainable development using the flow of funds provided by developed (or donor) countries by increasing the resilience of their environmental, social and economic systems to either endogenous or exogenous shocks. After a description of the observed data, there will follow a construction of a composite indicator capable of providing a quantitative measure of the environmental performance of a country related to other variables that are able to describe social and economic particularities. In this way, it is possible to individuate a way capable of giving a measure of the repercussion of the financial flows on combating environmental degradation seen through the most important components of the increase in global warming. Our results contribute to the debate on the vulnerability and resilience of receiving countries as part of the UN Framework Convention on Climate Change agreement.

Keywords Greenhouse gas • Environmental pollution index • Quantile regression • Clustered data • AidData • Energy generation • Biosphere protection

5.1 Data

Our aim is to analyze the determinants of countries' environmental performance and the effectiveness of climate funds. In other words, we want to investigate whether countries can progress towards more environmentally sustainable development using the flow of funds provided by donor countries and by increasing the resilience of their environmental, social and economic systems to either endogenous or exogenous shocks. Our results contribute to the debate on the vulnerability and resilience of receiving countries as part of the UN Framework Convention on Climate Change agreement.

As our approach is devoted to sketching the features of the receiving countries' development and to exploring the factors behind environmental performance, our empirical analysis employs a wide range of control variables. Variables can be grouped into four homogeneous areas: funds, energy, demographics, and socio-economic living standards, while the outcome variable summarizes environmental aspects.

© The Author(s) 2018
A.A. Romano et al., *Climate Finance as an Instrument to Promote the Green
Growth in Developing Countries*, SpringerBriefs in Climate Studies,
DOI 10.1007/978-3-319-60711-5_5

Definitions, data sources and descriptive statistics of variables for the sample (149 countries: treated and untreated) are shown in Table 5.1.

In the funds group we include the amount of aid from donor to recipient countries. The total amount of aid was retrieved from the AidData database (Tierney et al. 2011), a project that collects financial flows destined to aid developing countries. The funds received mainly include two types of aid. We analyze the flows destined to energy generation and supply with regard to power generation and renewable sources; these include funds for policy, planning, development programs, surveys and incentives to develop clean energy in the country. We consider the flows of funds targeted at (i) the general environment and (ii) biosphere protection (air pollution control, ozone layer preservation and marine pollution control). In general, both types of financial flows are among the aid directed to climate change, and these funds are for both climate adaptation and climate mitigation.

The 'energy' group refers to the class of electricity generation factors. It includes the share of non-hydroelectric (sh_nonhydro) and fossil (sh_fossil) generation expressed, respectively, as the ratio of non-hydroelectric generation to total electricity production and the ratio of fossil fuel electricity generation to total electricity production (see, e.g., Romano and Scandurra 2016).

In this group, we also include energy intensity. As suggested by Romano et al. (2015), more developed economies are also oriented to production efficiency improvement and low energy intensity and, for these reasons, the ratio between energy consumption and GDP can be considered as a proxy of technological and economic progress. In the same class of factors, we include the excavation of energy resources. In particular, we investigate the effects of oil supply, which includes the production of crude oil (including lease condensate), natural gas plant liquids and other liquids, and refinery processing gain.[1] With this indicator, we can control for lobbying effects (see, e.g., Marques et al. 2011; Marques and Fuinhas 2012). Where these resources are used intensively, we expect a lower usage of RES.

Among demographic characteristics, we include the percentage of the female population. Population is based on the de facto definition, which counts all residents regardless of legal status or citizenship – except for refugees not permanently settled in the country of asylum, who are generally considered part of the population of their country of origin. The share of the female population has been inserted as a proxy for preferences for greener policy management. It has been shown, in fact, that women have stronger preferences for environmental issues and protection (Zhao et al. 2012; Romano et al. 2016).

In the 'socio-economic and living standards' group, we include Gross Domestic Product (in logarithmic scale) to control for the relative level of economic development. In fact, it is commonly assumed that richer countries are able to better promote investments in renewable energy sources (Romano et al. 2015) and to

---

[1]Negative refinery processing gain data values indicate a net refinery processing loss.

Table 5.1 Data: Definitions, descriptive statistics and sources

Variable	Definition	Unit	Obs	Mean	Std. Dev.	Min	Max	Source
Funds								
Tot_rec	Sum of funds destined to "energy generation and supply" and to "general environmental protection"	M $	149	$1110	$4714	$ 0	$41,710	AidData.org
tot_rec_en	Funds destined to "energy generation and supply": Power generation/renewable sources: Including policy, planning, development programmes, surveys and incentives		149	$620	$2923	$ 0	$30,814	
tot_rec_bio	Funds destined to "general environmental protection": Biosphere protection: Air pollution control, ozone layer preservation; marine pollution control		149	$490	$3504	$ 0	$41,605	
Energy								
ei	Energy intensity using purchasing power parities is calculated by dividing the data on total primary energy consumption in quadrillion British thermal units for each country and year by the gross domestic product using purchasing power parities in billions of (2005) U.S. dollars for each available country and year.	Btu per year 2005 U.S. dollars (PPP)	149	6759	6215	199	50,976	The U.S. Energy Information Administration (EIA)
oil_sup	Total oil supply includes the production of crude oil (including lease condensate), natural gas plant liquids, and other liquids, and refinery processing gain.	Thousand barrels per day	149	464	1429	−0.54	10,908	
sh_foss	Fossil fuels electricity generation consists of electricity generated from coal, petroleum, and natural gas.	Billion kilowatt-hours	149	0.66	0.34	0	1	
sh_nonhydro	Hydroelectric generation excludes generation from hydroelectric pumped storage, where separately reported.		149	0.02	0.06	0	0.42	

(continued)

Table 5.1 (continued)

Variable	Definition	Unit	Obs	Mean	Std. Dev.	Min	Max	Source
Demographic								
pop_fem	Female population is the percentage of the population that is female. Population is based on the de facto definition of population, which counts all residents regardless of legal status or citizenship--except for refugees not permanently settled in the country of asylum, who are generally considered part of the population of the country of origin.	% of total	149	49.74	3.48	24.65	54.31	World Bank (world development indicators)
Socio-economic and living standards								
acc_el	Access to electricity is the percentage of population with access to electricity. Electrification data are collected from industry, national surveys and international sources.	% of population	149	72.21	32.55	3.5	100	World Bank (world development indicators)
lgdp	GDP per capita based on purchasing power parity (PPP). PPP GDP is gross domestic product converted to international dollars using purchasing power parity rates. An international dollar has the same purchasing power over GDP as the U.S. dollar has in the United States.	Constant 2011 international $	149	8.83	1.13	6.55	11.76	
ln_elcons	The electric consumption is the electric power consumption equal to the sum of total net electricity generation and electricity imports net of the electricity exports and electricity transmission and distribution. Losses	Billion kilowatthours	149	1.55	2.36	−3.84	8.24	The U.S. Energy Information Administration (EIA)

The table summarizes the variables used in the analysis, the measurement scale employed and the original data sources improve environmental conditions. Moreover, GDP is also related to energy consumption, which is considered a proxy for a country's economic development (Toklu et al. 2010). The countries included in Annex-II of the Kyoto Protocol are required to contribute financial resources to enable developing countries to undertake emissions reduction activities under the Convention and to help them adapt to the adverse effects of climate change. In addition, they have to "take all practicable steps" to promote the development and transfer of environmentally friendly technologies to economies in transition (EIT) Parties and developing countries. Funding provided by these countries is channeled mostly through the UNFCCC's financial mechanism.

We also include access to electricity expressed as the percentage of the population with direct access. Access to electricity is essential for social, economic, and political development (see, e.g., Kanagawa and Nakata 2007, 2008, Onyeji et al. 2012). Despite the enormous potential of most of the developing countries in fossil and renewable energy sources, however, some of these countries still suffer from major energy deficits. For this reason, the World Bank supports policies aimed at relieving energy poverty and thus improving the living conditions of the population in developing countries.

As an outcome variable, we built a composite indicator to capture environmental degradation due to economic development. The composite indicator proposed is an index that measures the environmental performance of countries and summarizes the effect of anthropogenic activities on the atmosphere. We consider the most important anthropogenic emissions of greenhouse gases (GHG): Carbon Dioxide (CO_2), Methane (CH_4), Nitrous oxide (N_2O) and Chlorofluorocarbons (CFCs), hydro chlorofluorocarbons (HCFCs), hydrofluorocarbons (HFCs), perfluorocarbons (PFCs), and sulfur hexafluoride (SF_6), together called F-gases. The concentration of GHG in the atmosphere is increasing and is the main cause of the greenhouse effect, mostly due to the burning of fossil fuels for human activities. All these are expressed in CO_2-equivalents.

5.2 Research Hypotheses

To analyze the expected divergences, we split the sample into two different sub-samples, classified into treated and untreated countries, with reference to countries that received (treated) or did not receive (untreated) financial aid.

Assumption 1. The Treated Countries Present Higher Energy Consumption than the Untreated Countries Several studies (Thomas et al. 2000; Aflaki et al. 2014) demonstrate that electricity consumption is the greatest contributor to the production of GHG emissions. To prove this assumption, we test electricity consumption (in logarithmic scale) to verify that it is higher for the treated countries than the countries that have not received climate funds. To assess whether the differences in electricity consumption between groups of countries are significantly different from

Table 5.2 Hypoothesis test for difference in mean in electricity consumption and share of fossil generation between receiving and not receiving countries

Two-sample t test					
Country groups	obs	Mean sh_foss	t-test	Mean lnelcons	t-test
Receiving	83	0.595	−2.538**	2.214	4.041***
Not receiving	66	0.735		0.717	

Notes: *: Statistical significance at 10%; **: Statistical significance at 5%; ***: Statistical significance at 1%.

zero, we test the hypothesis on the difference between the two population means (Table 5.2). Because there is significant evidence against equality in the means, we can confirm that the countries that have received aid to improve their energy management are more polluting and less rich than the untreated countries, and they require a large amount of electricity as well.

Assumption 2. Renewable Electricity Generation is Higher in Treated than Untreated Countries Sam et al. (2016) argue that developing countries need to grow their incomes quickly and prefer to allocate their resources to the least expensive technologies that utilize the traditional fuels. We test whether green funds have contributed to increasing the share of renewable generation in recipient countries and decreasing the share of fossil fuel generation. To assess if the differences in the shares of traditional sources in the groups of countries are significantly different from zero, we test the hypothesis on the difference between the two population means, using a simple t-test (Table 5.2). Because there is significant evidence against equality in the means, we can confirm that the receiving countries have a lower dependence on fossil fuel.

These assumptions confirm that the flows of funds are directed, primarily, toward countries with increasing electricity consumption. Furthermore, these countries present a lower share of fossil fuel generation. These indications suggest that climate funds are, at this stage, addressed to rapidly developing countries. For these reasons, we want to investigate the effectiveness of climate funds on environmental and energy projects and investigate the flow of funds among countries.

5.3   A Composite Indicator for Environmental Performance

This section is focused on an application of the proposed methodology to all countries (donor, treated and untreated) included in our dataset (176 countries), to propose an Environmental Pollution Index (EPI).

Considering the multi-dimensionality of the EPI, we aggregate the four main GHG.[2] To weigh the indicator we follow two different schemes: equal weighting and PCA.

According to Saisana and Saltelli (2011), the presence of few indicators justifies the use of an equal weighting scheme. In other words, the weight on each single indicator for each group of countries is equal to 0.25, according to Eq. (2). However, considering the strong correlation of the indicators, we also perform a PCA. One characteristic of the weighting procedure based on PCA is the two-step approach. In the first step, we have to aggregate the normalized indicators in the sub component index using the appropriate weight and, in the second step, the sub-indexes are aggregated to obtain the composite indicator.[3] For a more detailed discussion see, e.g., OECD (2008).

The calculation of the EPI has been performed following a geometric aggregation. The EPI obtained following the equal weighting schemes is in the Appendix. In Fig. 5.1, we report the graphical representation of the environmental indexes obtained.

Based on the methodology described in Sect. 3.1, the selected indicators, which quantify the concept of environmental performance in countries, will be employed for the construction of the composite indicator for countries for the year 2012. This indicator (i) captures the interactions and interdependencies of the selected indicators and (ii) facilitates the comparability of countries based on their index.

We observe that the two indexes follow a common path and, considering the countries' rankings, we do not observe changes in the positions occupied by countries, confirming the robustness of the indicator proposed.

Analyzing the results obtained, in the next part of the paper we refer to the EPI that is based on the equal weighting scheme.

The EPI is between zero (China and the United States) and 0.9999 (Kiribati). Higher values of the index indicate countries that have a lower level of GHG emissions. This does not imply that these countries have paid great attention to environmental problems. Most of the countries with higher EPI levels are generally underdeveloped, and they receive mitigation funds to promote their economic growth and development along a green path in which investments are concentrated on green electricity generation plants and supporting low-emission transport. These countries use financial resources to improve the living conditions of their populations and have a low share of generation from renewable sources. In poor countries, scarce resources are unlikely to be invested in technologies that do not meet high expectations (Mulugetta 2008; Bhattacharyya 2013; Desjardins et al. 2014).

---

[2]CO_2, CH_4, N_2O and F-gases.

[3]The elaborations are based on the "Compind" R package (Vidoli and Fusco 2015) for R (Core Team 2016).

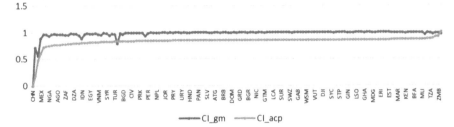

Fig. 5.1 Comparison between environmental pollution index obtained with equal (CI_gm) and PCA (CI_acp) weighting schemes

Table 5.3 Hypothesis test for difference in means of Environmental Pollution Index (EPI) ibetween receiving and not receiving countries

Two-sample t test			
Variable	Obs	Mean EPI	t-test
Receiving	83	0.963	-1.855^{a}
Not receiving	66	0.989	
Total	149		

H0: diff = mean(T) – mean(U) = 0
[a]: Denote statistical significance at 10%

On the opposite side, we observe developed and developing countries that have high or increasing economic and social development and contribute to the deterioration of the quality of the environment.

These results suggest that we may be able to prove a further assumption:

Assumption 3. Countries Receiving Climate Funds (Treated) Present Higher GHG Emissions than Untreated Countries As presented earlier, to assess whether the differences in GHG emissions between groups of countries are significantly different, we conducted a simple t-test whose null hypothesis is that there are no differences in the EPI means among groups, against the alternative that not all the means are equal (Table 5.3).

There is significant evidence against equality in the means and, therefore, this test confirms our last assumption: treated countries are more polluting compared to the countries that have not received climate funds.

This difference highlights that divergences of environmental characteristics exist between the two groups of countries.

5.4 Distribution of Climate Funds: A Way to Combat Environmental Degradation?

In this subsection, we discuss the key factors of the EPI, the divergences based on the environmental impact of each group of countries analyzed, and the effectiveness of the flow of funds received by developing countries.

To this end, we focus on countries that have received (treated) or not received (untreated) financial aid from the Annex-II (donor) countries. In other words, we remove the donor countries from the sample because they do not receive any environmental or energy benefits but, at the same time, they contribute to combatting climate change in developing countries (following UNFCCC agreements) and are also the major polluters.

Figures 5.2a and b present the histograms of the EPI for countries that receive climate funds (treated), countries that are not included in Annex-II, countries that have received financial aid (treated), and untreated countries. A first remark concerns the high values of the Environmental Pollution Index for the untreated countries, which therefore do not have a significant impact on climate change.

Fig. 5.2 Histograms of Environmental Pollution Index (EPI) in country groups

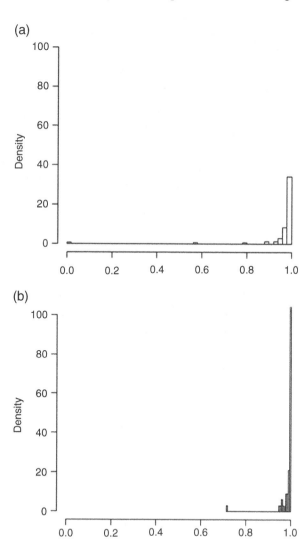

This is in contrast to the treated countries, which show lower levels of the index and an increase in heterogeneity due to the presence of large polluters such as China and India. This confirms that climate finance is mainly directed towards large-scale polluters. The EPI, particularly, is characterized by much higher values in the upper-half of its distribution. Due to the high skewness of the EPI distribution, the analysis is conducted by means of a quantile regression by focusing on the first quantiles (Neumayer et al. 2014) and on the difference between the aid directed to Biosphere Protection and that intended to promote Clean Energy according to AidData (Tierney et al. 2011).

To analyze the effectiveness of the climate funds, we estimate three different models considering (i) biosphere protection, (ii) green energy generation and (iii) total funds (the sum between biosphere and green energy) and by controlling for a set of variables. In this way, we test the usefulness of climate finance to promote green growth in developing countries.

The three estimated models are the following:

(a)
$$Q_{EPI}(\alpha|X_{gi}) = \begin{aligned}[t] &\beta_0(\alpha) + \beta_1 tot\_rec\_bio(\alpha) + \beta_2 pop\_fem(\alpha) \\ &+\beta_3 ei(\alpha) + \beta_4 oil\_sup(\alpha) + \beta_5 sh\_foss(\alpha) + \beta_6 sh\_nonhydro(\alpha) \\ &+\beta_7 lgdp(\alpha) + \beta_8 lnel\_cons(\alpha) + \beta_9 acc\_el(\alpha) \\ &+u(\alpha) \end{aligned}$$

(b)
$$Q_{EPI}(\alpha|X_{gi}) = \begin{aligned}[t] &\beta_0(\alpha) + \beta_1 tot\_rec\_en(\alpha) + \beta_2 pop\_fem(\alpha) \\ &+\beta_3 ei(\alpha) + \beta_4 oil\_sup(\alpha) + \beta_5 sh\_foss(\alpha) + \beta_6 sh\_nonhydro(\alpha) \\ &+\beta_7 lgdp(\alpha) + \beta_8 lnel\_cons(\alpha) + \beta_9 acc\_el(\alpha) \\ &+u(\alpha) \end{aligned}$$

(c)
$$Q_{EPI}(\alpha|X_{gi}) = \begin{aligned}[t] &\beta_0(\alpha) + \beta_1 Tot\_rec(\alpha) + \beta_2 pop\_fem(\alpha) \\ &+\beta_3 ei(\alpha) + \beta_4 oil\_sup(\alpha) + \beta_5 sh\_foss(\alpha) + \beta_6 sh\_nonhydro(\alpha) \\ &+\beta_7 lgdp(\alpha) + \beta_8 lnel\_cons(\alpha) + \beta_9 acc\_el(\alpha) \\ &+u(\alpha) \end{aligned}$$

where α shows the α-th quantile and $0 < \alpha < 1$, β_0 is the intercept, and β_1 to β_9 are the slopes of the independent variables. The three proposed models differ only in the finance variable. In the first we consider biosphere protection (Tot_rec_bio), in the second we include green energy generation (Tot_rec_en), and in the last model we assess the sum between the previous funds (Tot_rec). All the control variables are the same in the three models. In particular, *pop_fem* is the percentage of the female population; *ei* is the energy intensity; *oil supply* includes the production of crude oil, natural gas plant liquids and other liquids, and refinery processing gain; *sh_nonhydro* is the ratio of non-hydroelectric generation to total electricity production while *sh_fossil* is the fossil generation expressed as the ratio of fossil fuel generation to total electricity production; *lngdp* is Gross Domestic Product (in logarithmic scale) expressed in 2011 US\$ PPP; *lnel_cons* is the energy consumption (in logarithmic scale) while *acc_el* is the access to electricity. Finally, *u* is the error term that allows the disturbance conditionally to be independent across clusters but can be correlated within clusters to vary. The constant β_0 and the coefficients β_1 to β_9 are estimated for different quantiles ($\alpha = 0.1,\ldots, 0.95$) using the entire dataset each time. Moreover, to account for the presence of clusters in the observed countries, we conduct a quantile regression that estimates the robust covariance matrix with within-cluster dependence (Parente and Santos Silva 2016). To assess the presence of the clustered data, we conduct the tests of Machado- Santos Silva and Parente- Santos Silva (Parente and Santos Silva 2016). The results show that without clustering the sample, the Machado- Santos Silva test for heteroscedasticity rejects the null hypothesis of constant variance, so there exists intra cluster correlation. By clustering the sample by Income Group variable[4] to account for the economic and social structure of the countries, we accept the null hypothesis of the Parente- Santos Silva test of no-intra-cluster correlation. Therefore, we have estimated three different models with the latter method to obtain confidence intervals for quantile regression estimators at any single quantile.

Tables 5.4 and 5.6 report the vector of coefficients estimated both by OLS and by quantile regression (for $\alpha = 0.10, 0.20, 0.30, 0.40, 0.50, 0.60, 0.70, 0.80, 0.90, 0.95$). Figures 5.3 *(a) – (j)* report the quantile coefficients plot.

In particular, Table 5.4 reports the estimated coefficients of the amount of received flow of funds (Tot_rec), Table 5.5 those directed to power generation and renewable sources (Tot_rec_en) while Table 5.6 reports the funds targeted at the biosphere protection (Tot_rec_bio).

Accordingly, each of the columns (10th, 20th, ..., 95th) presents a vector of estimated coefficients in the regression for the corresponding quantile of EPI distribution, while the first column shows the estimates for the OLS regression.

Focusing on Table 5.4, we observe that the total flow of funds presents a significant relationship with the proposed environmental index. Estimated coefficients on control variables are generally significant and in line with our hypotheses.

The estimated coefficients for the total flow of funds (for total energy aid and for total biosphere protection) are higher in magnitude for the lower quantiles. This confirms the intuition that the countries with a lower level of EPI (generally the treated countries) are also the ones that receive more funds, while an increase in the EPI leads to a potential reduction in the received funds.

The estimated coefficients of the control variables, as shares of fossil fuels or the oil supply, also decrease with the increasing quantiles of the EPI, and the relation is stronger for the first quantiles. On the other hand, although the relation of energy intensity is positive for the bottom-half of the EPI distribution, it becomes negative for the central quantiles, and it becomes positive for the top-half distribution.

---

[4]We follow the income group distribution proposed by the World Bank (https://datahelpdesk. worldbank.org/knowledgebase/articles/906519-world-bank-country-and-lending-groups)

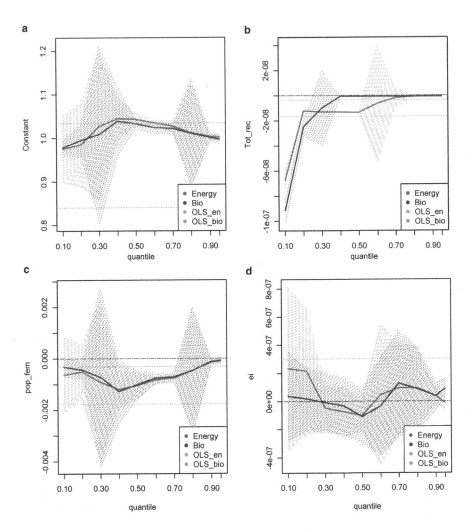

Fig. 5.3 (a) – (j) Quantile coefficient plots
The plots show the estimated coefficients for all different quantiles for the two estimated quantile regression model that involves flow of funds for the biosphere protection and those destined to power generation and renewable sources, respectively. The figures from *a* to *j* present, respectively, the coefficients $\beta0(\alpha)$ (Constant), $\beta1(\alpha)$ (Amount of funds), $\beta2(\alpha)$ (% of female population), $\beta3(\alpha)$ (energy intensity), $\beta4(\alpha)$ (oil supply), $\beta5(\alpha)$ (share of fossil generation), $\beta6(\alpha)$ (share of non-hydroelectric generation), $\beta7(\alpha)$ (Gross Domestic Product (in logarithmic scale)), $\beta8(\alpha)$ (energy consumption (in logarithmic scale)) and $\beta9(\alpha)$ (% access to electricity) for 99 different quantiles ($\alpha \in \{0.01, \ldots 0.99\}$) for the full regression model. The respective values are connected as a the *dashed line* (for the biosphere protection flow) and *dotted line* (power generation and renewable sources funds); the *grey* shading indicates the 95th point-wise confidence intervals about the coefficients, with the least squares result added as a horizontal *dotted* and *dashed line*, respectively. Note that there is an additional solid line at zero

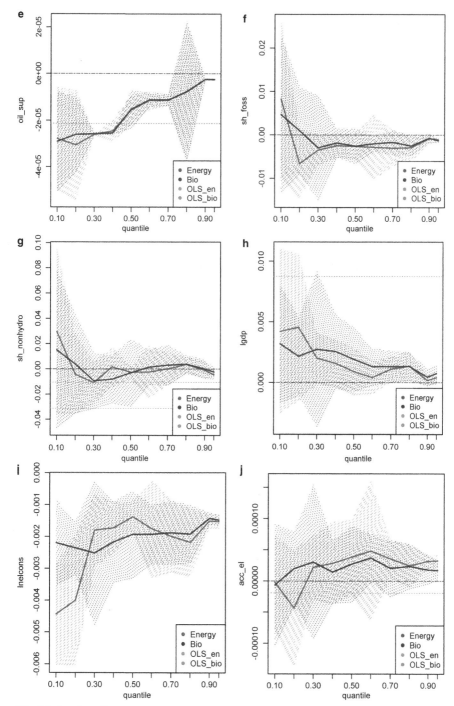

Fig. 5.3 (continued)

Table 5.4 OLS and quantile regression results for total disbursements

	OLS	10th	20th	30th	40th	50th	60th	70th	80th	90th	95th
	Coef./se	Coef./se	Coef./se	Coef./se	Coef./se	Coef./se	Coef./se	Coef./se	Coef./se	Coef./se	Coef./se
Tot_rec	−8.74e-09*** (1.65e-09)	−3.88e-08 (2.70e-08)	−1.85e-08*** (6.19e-09)	−1.29e-08*** (1.76e-10)	−1.30e-08*** (1.25e-10)	−1.69e-09 (4.92e-09)	−7.21e-10*** (1.13e-10)	−7.33e-10*** (9.64e-11)	−7.23e-10*** (7.59e-11)	−7.85e-10*** (3.96e-11)	−7.60e-10*** (1.19e-11)
pop_fem	−0.000902 (0.00130)	−0.000384 (0.000961)	−0.000350*** (0.000132)	−0.00106 (0.000668)	−0.00124*** (0.000366)	−0.00104*** (0.000260)	−0.000787*** (0.000102)	−0.000666** (0.000306)	−0.000460 (0.00134)	−0.000141*** (0.0000393)	−0.000105* (0.0000543)
ei	0.0000000223 (0.000000107)	8.64e-09 (0.000000335)	0.000000153 (0.000000153)	−6.62e-08 (9.46e-08)	−9.65e-08 (8.24e-08)	−0.000000115* (5.99e-08)	4.96e-08 (0.000000115)	0.000000115 (0.000000166)	8.82e-08 (0.000000156)	3.25e-08 (3.11e-08)	−2.84e-09 (2.82e-08)
oil_sup	−0.0000229 (0.0000146)	−0.0000288 (0.000139)	−0.0000289*** (0.000000666)	−0.0000261*** (0.000000233)	−0.0000252*** (0.00000222)	−0.0000153*** (0.00000389)	−0.0000115** (0.00000123)	−0.0000105** (0.00000446)	−0.00000741 (0.000000741)	−0.00000243*** (0.000000182)	−0.00000245*** (7.09e-08)
sh_foss	−0.0180* (0.00698)	0.0156* (0.00632)	−0.00219 (0.00225)	−0.00257*** (0.000678)	−0.00204** (0.000823)	−0.00247 (0.00166)	−0.00213 (0.00150)	−0.00204 (0.00181)	−0.00290** (0.00132)	−0.000833* (0.000503)	−0.00109*** (0.000230)
sh_nonhydro	−0.0331 (0.0475)	0.0379 (0.0508)	−0.00468 (0.00764)	−0.0102 (0.00751)	−0.00659 (0.00492)	−0.00471 (0.0128)	0.000214 (0.00846)	0.00158 (0.00701)	0.00301 (0.00454)	−0.00122 (0.00152)	−0.00383** (0.00175)
lgdp	0.0153 (0.0150)	0.00209 (0.00448)	0.00206* (0.00108)	0.00189 (0.00133)	0.00133 (0.000877)	0.00130 (0.000995)	0.000883 (0.000209)	0.00113* (0.000585)	0.00128 (0.000209)	0.000142 (0.000209)	0.000386 (0.000357)
lnelcons	−0.000906** (0.00259)	−0.00201* (0.00107)	−0.00190** (0.000842)	−0.00168*** (0.000313)	−0.00138*** (0.000234)	−0.00169*** (0.000274)	−0.00170*** (0.000563)	−0.00179*** (0.000593)	−0.00210*** (0.000592)	−0.00145*** (0.000203)	−0.00145*** (0.0000825)
acc_el	−0.0000986 (0.000267)	−0.000000340 (0.0000932)	0.000000410 (0.0000268)	0.0000301 (0.0000313)	0.0000253 (0.0000303)	0.0000377 (0.0000305)	0.0000318* (0.0000187)	0.0000256 (0.0000167)	0.0000264 (0.0000175)	0.0000286*** (0.00000674)	0.0000286*** (0.00000674)
Constant	0.937*** (0.170)	0.981*** (0.0247)	0.9997*** (0.0142)	1.034*** (0.0417)	1.049*** (0.0228)	1.038*** (0.00975)	1.029*** (0.00825)	1.022*** (0.0155)	1.012*** (0.0638)	1.004*** (0.00272)	1.001*** (0.00524)
R-squared	0.482	0.377	0.436	0.452	0.449	0.369	0.324	0.337	0.378	0.459	0.459
N. of cases	149	149	149	149	149	149	149	149	149	149	149
Parente-Santos Silva test p-value		0.460	0.411	0.253	0.237	0.628	0.341	0.294	0.159	0.853	0.0719

The dependent variable is EPI.

The quantile regression estimations are made by the Stata's qreg2 command that reports standard errors and t-statistics that are asymptotically valid under heteroskedasticity and misspecification. Heteroskedasticity-robust standard errors are in parenthesis.

* Denote statistical significance at 10%.

** Denote statistical significance at 5%.

*** Denote statistical significance at 1%.

Table 5.5 OLS and quantile regression results for energy aid

	OLS	10th	20th	30th	40th	50th	60th	70th	80th	90th	95th
	Coef./se	Coef./se	Coef./se	Coef./se	Coef./se	Coef./se	Coef./se	Coef./se	Coef./se	Coef./se	Coef./se
tot_rec_en	-1.66e-08*	-6.71e-08***	-1.22e-08***	-1.30e-08***	-1.31e-08***	-1.35e-08***	-6.25e-09	-1.86e-09	-7.74e-10**	-7.14e-10***	-7.85e-10***
	(7.74e-09)	(3.77e-09)	(2.31e-09)	(2.66e-10)	(1.31e-10)	(9.58e-11)	(2.39e-08)	(3.58e-09)	(3.44e-10)	(1.59e-10)	(5.99e-11)
pop_fem	-0.00177***	-0.000629	-0.000510	-0.000911	-0.00120**	-0.00105***	-0.000823***	-0.000755***	-0.000470	-0.000144***	-0.000100**
	(0.000381)	(0.000745)	(0.000550)	(0.00161)	(0.000389)	(0.000268)	(0.0000859)	(0.0000884)	(0.00126)	(0.0000489)	(0.0000418)
ci	0.000000302**	0.000000233	0.000000217	-4.73e-08	-7.41e-08	-0.000000101*	4.65e-08	9.45e-08	9.39e-08	3.65e-08	-1.22e-08
	(7.30e-08)	(0.000000295)	(0.000000230)	(7.96e-08)	(9.04e-08)	(5.71e-08)	(0.000000163)	(0.000000163)	(0.000000148)	(2.94e-08)	(2.45e-08)
oil_sup	-0.0000214	-0.0000279**	-0.0000306**	-0.0000261***	-0.0000246***	-0.0000155***	-0.0000115***	-0.0000114***	-0.00000753	-0.0000241***	-0.0000244***
	(0.0000126)	(0.0000111)	(0.0000120)	(0.000000209)	(0.00000233)	(0.00000317)	(0.00000198)	(0.00000116)	(0.00000152)	(0.000000123)	(7.65e-08)
sh_foss	-0.0207	0.00826	-0.00662	-0.00351**	-0.00248*	-0.00265*	-0.00286	-0.00305	-0.00294*	-0.000858	-0.00108***
	(0.0126)	(0.00901)	(0.00409)	(0.00173)	(0.00139)	(0.00145)	(0.00207)	(0.00207)	(0.00156)	(0.000612)	(0.000157)
sh_nonhydro	-0.0312	0.0298	-0.00430	-0.0108	0.00162	-0.00259	-0.00246	0.0000431	0.00359	-0.000925	-0.00452**
	(0.0521)	(0.0330)	(0.0159)	(0.00827)	(0.00803)	(0.00235)	(0.0105)	(0.00628)	(0.00324)	(0.00346)	(0.00184)
lgdp	0.00876	0.00422	0.00455	0.00200*	0.00158	0.000906	0.000410	0.00110**	0.00135**	0.000174	0.000405
	(0.00652)	(0.00341)	(0.00307)	(0.00106)	(0.000982)	(0.000981)	(0.000685)	(0.000529)	(0.000629)	(0.000198)	(0.000278)
lnelcons	-0.00910***	-0.00444***	-0.00400***	-0.00180***	-0.00173***	-0.00138***	-0.00175***	-0.00199***	-0.00218***	-0.00151***	-0.00153***
	(0.00181)	(0.000802)	(0.00103)	(0.000578)	(0.000403)	(0.000231)	(0.000778)	(0.000569)	(0.000538)	(0.000117)	(0.000110)
acc_el	-0.0000210	-0.000000192	-0.0000433	0.0000221	0.0000286	0.0000385	0.0000482	0.0000358***	0.0000246	0.0000319***	0.0000326***
	(0.000156)	(0.0000382)	(0.0000472)	(0.0000260)	(0.0000335)	(0.0000301)	(0.0000567)	(0.0000158)	(0.0000175)	(0.00000621)	(0.00000441)
Constant	1.034***	0.975***	0.985***	1.027***	1.045***	1.043***	1.035***	1.027***	1.012***	1.004***	1.001***
	(0.0605)	(0.0396)	(0.0486)	(0.0859)	(0.0247)	(0.00853)	(0.00272)	(0.00682)	(0.0644)	(0.00328)	(0.00400)
R-squared	0.551	0.459	0.504	0.518	0.523	0.528	0.532	0.394	0.361	0.453	0.462
N. of cases	149	149	149	149	149	149	149	149	149	149	149
Parente-Santos Silva test p-value		0.853	0.799	0.215	0.237	0.226	0.267	0.749	0.186	0.451	0.188

The dependent variable is EPI.

The quantile regression estimations are made by the Stata's qreg2 command that reports standard errors and t-statistics that are asymptotically valid under heteroskedasticity and misspecification. Heteroscedasticity-robust standard errors are in parenthesis.

* Denote statistical significance at 10%.

** Denote statistical significance at 5%.

*** Denote statistical significance at 1%.

Table 5.6 OLS and quantile regression results for biosphere protection aid

	OLS	10th	20th	30th	40th	50th	60th	70th	80th	90th	95th
	Coef./se	Coef./se	Coef./se	Coef./se	Coef./se	Coef./se	Coef./se	Coef./se	Coef./se	Coef./se	Coef./se
tot_rec_bio	-3.60e-09***	-8.87e-08***	-2.54e-08***	-1.02e-08	-6.30e-10***	-6.93e-10***	-7.08e-10***	-7.18e-10***	-7.36e-10***	-7.89e-10***	-8.02e-10***
	(3.40e-10)	(5.09e-09)	(4.61e-09)	(1.40e-08)	(1.58e-10)	(1.45e-10)	(9.90e-11)	(8.76e-11)	(5.65e-11)	(1.18e-11)	(1.33e-11)
pop_fem	-0.000309	-0.000400*	-0.000425**	-0.000745	-0.00128**	-0.00101***	-0.000755***	-0.000714***	-0.000470	-0.000124**	-0.0000774
	(0.00171)	(0.000213)	(0.000208)	(0.00164)	(0.000611)	(0.000284)	(0.000104)	(0.0000613)	(0.000470)	(0.0000509)	(0.0000565)
ei	0.000000106**	-3.19e-08	8.18e-09	-1.32e-08	-3.27e-08	-0.000000109	-3.42e-08	0.000000128	8.79e-08	3.70e-08	8.89e-08**
	(0.000000265)	(7.56e-08)	(8.78e-08)	(9.77e-08)	(0.000000101)	(9.98e-08)	(0.000000176)	(0.000000192)	(0.000000148)	(5.09e-08)	(3.54e-08)
oil_sup	-0.0000213	-0.0000304***	-0.0000261***	-0.0000258***	-0.0000255***	-0.0000152***	-0.0000112***	-0.0000112***	-0.00000777	-0.00000243***	-0.00000252***
	(0.0000153)	(0.0000113)	(0.00000803)	(0.000000258)	(0.000000294)	(0.00000411)	(0.00000149)	(0.00000122)	(0.0000149)	(0.000000147)	(6.30e-08)
sh_foss	-0.0265***	0.00542	0.00156	-0.00274	-0.00185	-0.00258	-0.00203**	-0.00169	-0.00255**	-0.000756	-0.00130***
	(0.00367)	(0.00827)	(0.00510)	(0.00520)	(0.00164)	(0.00204)	(0.000848)	(0.00146)	(0.00108)	(0.000476)	(0.000231)
sh_nonhydro	-0.0106	0.0161	0.00490	-0.00896	-0.00803	-0.00309	0.00138	0.00305	0.00384	0.0000179	-0.00235
	(0.0633)	(0.0250)	(0.0190)	(0.0108)	(0.00963)	(0.0135)	(0.00698)	(0.00812)	(0.00373)	(0.00361)	(0.00168)
lgdp	0.0242	0.00312*	0.00209	0.00264	0.00256*	0.00191	0.00131*	0.00132**	0.00133**	0.000445	0.000755*
	(0.0158)	(0.00158)	(0.00175)	(0.00263)	(0.00153)	(0.00153)	(0.000775)	(0.000650)	(0.000583)	(0.000360)	(0.000385)
lnelcons	-0.0157***	-0.00208***	-0.00224***	-0.00245***	-0.00217***	-0.00193***	-0.00193***	-0.00188***	-0.00192***	-0.00143***	-0.00148***
	(0.00204)	(0.000499)	(0.000511)	(0.000871)	(0.000644)	(0.000691)	(0.000472)	(0.000501)	(0.000463)	(0.0009900)	(0.0009913)
acc_el	-0.000232	-0.00000259	0.0000173	0.0000291	0.0000145	0.0000273	0.0000370*	0.0000206	0.0000237	0.0000179	0.0000171**
	(0.000316)	(0.0000488)	(0.0000329)	(0.0000507)	(0.0000253)	(0.0000309)	(0.0000198)	(0.0000230)	(0.0000183)	(0.0000114)	(0.00000771)
Constant	0.840**	0.982***	0.995***	1.011***	1.039***	1.033***	1.024***	1.022***	1.011***	1.001***	0.997***
	(0.192)	(0.0118)	(0.0216)	(0.0932)	(0.0401)	(0.00566)	(0.00674)	(0.00605)	(0.0604)	(0.00490)	(0.00548)
R-squared	0.336	0.0785	0.145	0.234	0.234	0.253	0.267	0.268	0.289	0.304	0.309
N. of cases	149	149	149	149	149	149	149	149	149	149	149
Parente-Santos Silva test p-value		0.679	0.241	0.314	0.478	0.244	0.253	0.491	0.385	0.394	0.722

The dependent variable is EPI.

The quantile regression estimations are made by the Stata's qreg2 command that reports standard errors and t-statistics that are asymptotically valid under heteroskedasticity and misspecification. Heteroscedasticity-robust standard errors are in parenthesis.

* Denote statistical significance at 10%.

** Denote statistical significance at 5%.

*** Denote statistical significance at 1%.

Therefore, almost all of the control variables are mean reverting while the flow of funds, oil supply, electricity consumption and GDP show a stronger relationship with the more polluting countries and with the first quantile of the EPI distribution.

In Tables 5.5 and 5.6, we report the estimated coefficients considering the funds received by countries for energy generation and biosphere protection, respectively.

In the energy framework, the estimated parameters are overall significant, with the exception of those concerning the share of non-hydroelectric generation and energy intensity.

In the biosphere protection framework, the significance in parameters is similar to the one listed for the energy funds. The exception is for electricity access, which becomes significant only for the last quantiles when considering the energy framework; they are still not significant with regard to air pollution control.

In all three estimated models, the coefficients of the flow of funds are overall significant and decrease as the quantiles increase. The difference between the flow of funds targeted at biosphere protection and those directed at power generation and renewable resources is due to the strengthening of the relationship between the first quantiles of the EPI and the related flow of funds. The relation is negative for all quantiles, but the estimated coefficients tend to zero for the higher quantiles.

Particularly, the estimated coefficients for the flow of funds directed to improve power generation have a negative relationship over all quantiles; after a rapid increase starting at quantile 0.2, the parameters tend to zero. This means that the flow of funds decreases when the EPI increases, in other words, when countries become cleaner and no longer need aid improve their energy production systems. However, these cleaner countries are also the least developed, and they have to adopt the new development pattern. For this reason, we observe that the funds provided by donor countries are mainly directed to and concentrated in countries with economic systems in transition, or to rapidly developing countries in order to combat climate change and promote green growth.

The total aid for biosphere protection shows two patterns: the former shows bottom distribution and the latter, after a rapid increase, from the 60th quantile becomes mean reverting around zero. This confirms that with the increase in the EPI, aid decreases, essentially removing the effects of the financial flows to these countries.

In the same manner, the shape of the oil supply coefficient across the three models is similar. The strength of the negative relation between the EPI and oil supply decreases gradually with the increase in EPI and reaches the nearly zero value at quantile 0.95. This is because the recipients of funds are developing countries and require more energy consumption (refer to exception to Kyoto protocol) to promote economic development; they justify the pattern of this latter variable that shows a negative relation over all quantiles. This relationship is stronger for the upper-half distribution of the EPI variable than for electricity access, for which the relation becomes positive and statistically significant only for the last quantiles. This is in line with the hypothesis that if developing countries can access the electrical grid they can also access renewable sources. For this reason, the positive relation between the percentage of the population with access to electricity and the EPI is because these populations will ultimately cease their use of traditional fossil fuels, improving the living conditions of rural developing areas (Kanagawa and Nakata 2007; Khennas 2012). Many of the funds targeted at climate change are also destined to increase this access through off-grid generation (Kaygusuz 2012).

As stated by Urmee and Md (2016), renewable sources can improve the living conditions and economic development of developing countries, but this process depends on the socio-economic factors that create environmental consciousness, such as the demographic structure and the level of education. Particularly, the role of women in developing countries has been studied with regard to household use of energy. Cecelski (2001) lists the main differences and threats related to energy engenderment between northern and southern women, focusing on the important educational role of the female population, while Martinot et al. (2002) argue that the living conditions of woman in the rural areas of developing countries are improving. The relationship, over all quantiles, between the EPI and the size of the female population is negative. This contrasts with the literature (Elnakat and Gomez 2015) that assesses the importance of women in developed countries.

Women play an important role in climate change, especially in the rural areas of the world. Many authors (Karekezi and Kithyoma 2002; Karki et al. 2005; Kanagawa and Nakata 2007; Walekhwa et al. 2009) argue that women, due to their responsibilities and functions in households, should be the main beneficiaries of aid and programs to improve living conditions (Rao and Reddy 2007).

The discounted negative relation cannot be seen as contrasting with the literature. It is possible that the absolute poverty of the considered countries and the lack of specific programs targeted at women do not allow this improvement.

The share of fossil fuel electricity generation and the share of non-hydroelectric generation seem to have the same pattern, even if the latter is statistically significant only for $\alpha = 0.95$ to both total energy aid and total biosphere protection aid. Pfeiffer and Mulder (2013) argue that non-hydroelectric electricity generation is slowed by aid and by the high use of fossil fuels and accelerated by supportive economic instruments. This can explain its non-significance as a determinant of EPI.

References

Aflaki S, Basher SA, Masini A (2014) Does economic growth matter? Technology-push, demand-pull and endogenous drivers of innovation in the renewable energy industry. HEC Paris Research Paper No. MOSI-2015-1070

Bhattacharyya SC (2013) Financing energy access and off-grid electrification: a review of status, options and challenges. Renew Sust Energ Rev 20:462–472

Cecelski E (2001) Gender perspectives on energy for CSD-9. Draft position paper including recommendations proposed by the ENERGIA Support Group and the CSD NGO Women's Caucus. Available at http://www.energia.org/resources/papers/csdposition.html

Desjardins S, Gomes R, Pursnani P, West C (2014) Accelerating access to energy: lessons learned from efforts to build inclusive energy markets in developing countries. http://www.

shellfoundation.org/ShellFoundation.org_new/media/Shell-Foundation-Reports/Access_to_ Energy_Report_2014.pdf. Accessed 14 Jan 2016

Elnakat A, Gomez JD (2015) Energy engenderment: an industrialized perspective assessing the importance of engaging women in residential energy consumption management. Energy Policy 82:166–177. doi:10.1016/j.enpol.2015.03.014

Kanagawa M, Nakata T (2007) Analysis of the energy access improvement and its socio-economic impacts in rural areas of developing countries. Ecol Econ 62(2):319–329

Kanagawa M, Nakata T (2008) Assessment of access to electricity and the socio-economic impacts in rural areas of developing countries. Energy Policy 36(6):2016–2029

Karekezi, S., & Kithyoma, W. (2002). Renewable energy strategies for rural Africa: is a PV-led renewable energy strategy the right approach for providing modern energy to the rural poor of sub-Saharan Africa? Energy Policy, 30(11–12), 1071–1086. doi:http://dx.doi.org/10.1016/ S0301-4215(02)00059-9

Karki SK, Mann MD, Salehfar H (2005) Energy and environment in the ASEAN: challenges and opportunities. Energy Policy 33(4):499–509. doi:http://dx.doi.org/10.1016/j.enpol.2003.08. 014

Kaygusuz, K. (2012). Energy for sustainable development: a case of developing countries. Renew Sust Energ Rev, 16(2), 1116–1126. doi:http://dx.doi.org/10.1016/j.rser.2011.11.013

Khennas S (2012) Understanding the political economy and key drivers of energy access in addressing national energy access priorities and policies: African perspective. Energy Policy 47(Supplement 1):21–26. doi:http://dx.doi.org/10.1016/j.enpol.2012.04.003

Marques AC, Fuinhas JA (2012) Are public policies towards renewables successful? Evidence from European countries. Renew Energy 44:109–118

Marques AC, Fuinhas JA, Manso JP (2011) A quantile approach to identify factors promoting renewable energy in European countries. Environ Resource Econ 49:351–366

Martinot E, Chaurey A, Lew D, Moreira JR, Wamukonya N (2002) Renewable energy markets in developing countries. Annu. Rev. Energy Environ 27:309–348

Mulugetta Y (2008) Human capacity and institutional development towards a sustainable energy future in Ethiopia. Renew Sust Energ Rev 12:1435–1450

Neumayer E, Plümper T, Barthel F (2014) The political economy of natural disaster damage. Glob Environ Chang 24:8–19. doi:http://dx.doi.org/10.1016/j.gloenvcha.2013.03.011

OECD (2008) Handbook on constructing composite indicators: methodology and user guide. OECD, Paris

Onyeji I, Bazilian M, Nussbaumer P (2012) Contextualizing electricity access in sub-Saharan Africa. Energy Sustain Dev 16(4):520–527

Parente PM, Santos Silva J (2016) Quantile regression with clustered data. J Econ Methods 5 (1):1–15. doi:10.1515/jem-2014-0011

Pfeiffer B, Mulder P (2013) Explaining the diffusion of renewable energy technology in developing countries. Energy Econ 40:285–296. doi:http://dx.doi.org/10.1016/j.eneco.2013.07.005

R Core Team (2016) R: a language and environment for statistical computing. R Foundation for Statistical Computing, Vienna. URL https://www.R-project.org/

Rao MN, Reddy BS (2007) Variations in energy use by Indian households: an analysis of micro level data. Energy 32(2):143–153. doi:http://dx.doi.org/10.1016/j.energy.2006.03.012

Romano AA, Scandurra G (2016) Divergences in the determinants of investments in renewable energy sources: hydroelectric vs. other renewable sources. J Appl Stat 43(13):2363–2376. doi:10.1080/02664763.2016.1163526

Romano AA, Scandurra G, Carfora A (2015) Probabilities to adopt feed in tariff conditioned to economic transition: a scenario analysis. Renew Energy 83:988–997

Romano AA, Scandurra G, Carfora A, Pansini RV (2016) Assessing the determinants of SIDS' pattern toward sustainability: a statistical analysis. Energy Policy. Forthcoming. doi:10.1016/j. enpol.2016.03.042

Saisana M, Saltelli A (2011) Rankings and ratings: instructions for use. Hague J Rule Law 3 (2):247–268

Sam A, Syed Abul B, Andrea M (2016) Does economic growth matter? Technology-push, demand-pull and endogenous drivers of innovation in the renewable energy industry. MPRA – Working paper. https://mpra.ub.uni-muenchen.de/69773/1/MPRA_paper_69773.pdf

Thomas C, Rolls J, Tennant T (2000) The GHG indicator: UNEP guidelines for calculating greenhouse gas emissions for businesses and non-commercial organisations. UNEP, Paris

Tierney MJ, Nielson DL, Hawkins DG, Timmons Roberts J, Findley MG, Powers RM, Parks B, Wilson SE, Hicks RL (2011) More dollars than sense: refining our knowledge of development finance using AidData. World Dev 39(11):1891–1906

Toklu E, Guney MS, Isik M et al (2010) Energy production, consumption, policies and recent developments in Turkey. Renew Sust Energ Rev 1:1172–1186

Urmee T, Md A (2016) Social, cultural and political dimensions of off-grid renewable energy programs in developing countries. Renew Energy 93:159–167. doi:10.1016/j.renene.2016.02.040

Vidoli, F., Fusco, E. (2015). Compind: composite indicators functions. R package version 1.1. https://CRAN.R-project.org/package=Compind

Walekhwa PN, Mugisha J, Drake L (2009) Biogas energy from family-sized digesters in Uganda: critical factors and policy implications. Energy Policy 37(7):2754–2762. doi:http://dx.doi.org/10.1016/j.enpol.2009.03.018

Zhao SX, Chan RC, Chan NYM (2012) Spatial polarization and dynamic pathways of foreign direct investment in China 1990–2009. Geoforum 43(4):836–850

Chapter 6
Conclusions and Policy Implications

Abstract In this chapter we report some concluding remarks and policy implications. The results obtained by the analysis of the committed and disbursed flow of funds revealed a strong heterogeneity in the way the funds are being allocated by donors; however, our findings show a relationship between the amounts disbursed and greenhouse gas emissions, which is more significant with regard to the funds directed toward biosphere protection.

Our study shows that close attention should be paid to the analysis of political contexts in a broad sense. Particularly, we must focus on the international negotiations process that enables the direction of funds toward specific needs and priorities and the issue of access to electricity. For example, the difficulties that developing countries face when trying to improve their green economic development without access to carbon remain a matter of the utmost importance and urgency for many developing countries that lack significant aid from developed countries.

Keywords Greenhouse gas • Clustered data • Flow of funds • Green economic development • Policy implications

Since 1992, the UNFCCC has set out a framework for international action to stabilize GHG emissions to prevent and combat climate change. The UNFCCC recognizes that developed countries have contributed the most to the global accumulation of GHG emissions, while developing countries bear less historical responsibility. This recognition has led to a commitment from developed countries to mobilize finance to help developing countries respond to climate change, and such 'climate finance' has become a central issue in international negotiations.

During the COP15 held in Copenhagen, the developed countries guaranteed immediate "Fast-start Finance" of up to USD 30 billion over 2010–2012 to launch the project and pledged to cooperatively provide USD 100 billion annually by 2020 to developing countries.

Climate finance want to support developing countries by enabling poorer states to move towards low-emission and climate-resilient development pathways. Particular attention was dedicated to vulnerable countries, including Small-Island

© The Author(s) 2018
A.A. Romano et al., *Climate Finance as an Instrument to Promote the Green Growth in Developing Countries*, SpringerBriefs in Climate Studies, DOI 10.1007/978-3-319-60711-5_6

Developing States (SIDSs), Least-Developed Countries (LDCs), and African States.

The aims of this work are to (*i*) demonstrate the existence of preferential channels in "Fast-start Finance" between developed and developing countries, (*ii*) examine the funds' distribution based on environmental pollution, and (*iii*) evaluate whether these preferences could undermine the effectiveness of such measures in combating worsening environmental conditions. To reach our aims, we develop a multi-step procedure using a large dataset of countries.

We consider the flow of funds directed to economic infrastructure, to services promoting the sustainable use of energy, and to specific measures to protect or enhance the physical environment. This study can be used to provide a basis for assessing the effectiveness of the flow of funds for environmental policy and laws, regulations, and economic instruments adopted in decision-making procedures.

The funds directed to "Fast-start Finance" meet the two requirements of ensuring the reduction in greenhouse gases and promoting the sustainable development of developing countries. The effectiveness of this very important tool is very difficult to determine and quantify due to several factors related to the socio-economic structures of developing countries, e.g., the difficulty developing countries face in creating and strengthening the power generation of electricity from RES without an adequate electricity grid, or, additionally, the difficulties they face in developing their economic systems without using carbon fossil sources and without the financial support of developed countries.

Developing countries' governments are rightly concerned about potential tensions between sustaining the economic growth needed to generate jobs and reduce poverty and reducing GHG emissions. As argued by Espagne (2016), these countries face the difficulty of improving their economic systems without access to carbon (due to high carbon prices) and without strong financial support from developed countries. Moreover, electricity access remains a key question for many developing countries, as argued by Bhattacharyya (2013). All this results in a lack of resilience, the importance of which becomes even more basic if we consider that the economic development of a country depends not only on the improved coordination of aid but also on the way aid is organized and distributed.

The initial assumptions concerning the existence of a significant correlation between the level of GHG emissions and the developing countries that receive funds has been directly verified by several tests, which also showed that these countries are more polluting and are energy-intensive "users". The increasing energy hunger of developing countries can only be met through the implementation of an adequate system of energy production, as well as management of the electricity transmission and distribution grid, which is almost completely absent in many areas of these countries.

The quantile regressions, on the other hand, confirm that there is a link between the Environmental Pollution Index, which represents the synthesis of GHG emissions, and climate finance. Particularly, our findings show that there is a relationship between the amounts disbursed and GHG emissions, which is more significant with regard to funds for biosphere protection. All three estimated models show that aid becomes mean, reverting around zero when the degree of EPI increases. In other words, the link between the oil supply, electricity consumption, and Environmental Pollution Index, as had been expected, is negative. Indeed, the more polluting developing countries are also energy-intensive "users". These countries receive more funds and highlight the presence of preferential channels with several donor countries.

The lack of a statistically significant relationship with the share of non-hydro renewable generation may seem discouraging. This may be due to the circumstance that, although the analyzed funds are directed to promote electricity generation by renewable sources, the reduction in greenhouse emissions can be considered an indirect effect of the creation of efficient energy production plants, which is observable, however, only after the passage of the number of years necessary to create an adequate system of energy production. The positive relationship of electricity access in the last quantile (where the countries are more "green") and GDP to the Environmental Pollution Index can perhaps support this theory in part, given that it is plausible to assume that a high level of GDP helps to secure access to the electricity grid in a sustainable way.

To evaluate, however, the real environmental effectiveness of climate finance, close attention should also be paid to the analysis of the political context in a broad sense. Several factors affect the achievement of the objectives of climate finance. These include the international negotiations process that enables funds to be directed towards specific needs and priorities, as well as access to electricity. For example, developing countries face major difficulties when trying to improve their economic systems without access to carbon; this remains a matter of the utmost importance and urgency for many developing countries without significant aid from developed countries.

The analysis of the committed and disbursed flow of funds known as "Fast-start Finance" revealed a strong heterogeneity in the way the funds are being allocated by donors. In particular, the European countries seem to focus on commercial partners and concentrate their funds in specific recipient countries, unlike the USA and Japan, which have allocated their aid (in lower amounts than some European countries) across various and different projects intended for many developing countries. To improve the effectiveness of climate funds, we suggest redesigning aid schemes not only to combat climate change but also to promote resilience to extreme events and reduce dependence on preferential channels with developed countries (which some recipients have). The funds can help to reach this goal by supporting the construction of an economically and environmentally sustainable framework, from the realization of off-grid electricity to improving relationships between political actors, which have become essential to achieving these important goals. Countries can indeed increase their energy sustainability, as academic reviews suggest. The sustainability and impact of RES generation largely depend on its suitability for potential end users. In these countries, the lack of electricity transmission networks requires the installation of off-grid RES power plants that could be financed by international organizations and governmental subsidies without inhibiting the development of commercial renewable energy technologies (RET) markets (Chaurey et al. 2012; Sovacool and Drupady 2012).

In general, the level of the EPI is low, and it can be improved with a transition to low carbon and climate resilient development that can be achieved through the promotion of RES generation, the improvement of energy efficiency, and the adoption of green policies to support clean economic growth (World Bank 2014).

As suggested by Bourguignon and Platteau (2015) and Pickering et al. (2015), a balanced and supranational organization needs to control, monitor and manage these funds to avoid the dispersion of the same funds, the futile duplication of projects, and the inevitable increase in the risks and costs of allocating them.

The heterogeneity in the distribution of funds makes it difficult to quantify the average effects of climate aid in each country in terms of the reduction in GHG emissions. Some other aspects of this problem would be interesting to investigate, for example, (1) the effect of climate finance on developing countries, distinguishing between treated and untreated countries; (2) the evaluation of the best way to address the funds that concentrate resources on a few specific projects (e.g., the effects of increasing and diversifying them); (3) the incidence of the time dimension in the distribution of funds. These points will be the foci of future studies.

References

Bhattacharyya SC (2013) Financing energy access and off-grid electrification: a review of status, options and challenges. Renew Sustain Energy Rev 20:462–472

Bourguignon F, Platteau JP (2015) The hard challenge of aid coordination. World Dev 69:86–97. doi:10.1016/j.worlddev.2013.12.011

Chaurey A, Krithika PR, Palit D, Rakesh S, Sovacool BK (2012) New partnerships and business models for facilitating energy access. Energy Policy 47(S1):48–55. doi:10.1016/j.enpol.2012.03.031

Espagne E (2016) Climate finance at COP21 and after: lessons learnt. CEPII, Policy Brief 2016–09, CEPII research center

Pickering J, Skovgaard J, Kim S, Roberts JT, Rossati D, Stadelmann M, Reich H (2015) Acting on climate finance pledges: inter-agency dynamics and relationships with aid in contributor states. World Dev 68:149–162

Sovacool BK, Drupady IM (2012) Energy access, poverty, and development. The governance of small-scale renewable energy in developing Asia. Ashgate, Farnham. https://www.book2look.com/embed/9781317143734 Accessed 12 Oct 2016

World Bank (2014) SIDS towards a sustainable energy future. WB-UN high level dialogue on advancing sustainable development in small island developing states

Appendices

Appendix A

Plots and Quantile Regression Using R and Stata Commands

Introduction

This appendix shows the basic commands to replicate the analysis, the tables and the plots proposed in the previous chapters using two statistical packages: R (R Core Team 2016). and Stata. R and Stata commands are shown in bold font while the comments using regular font and preceded by a "#" the hash character.

R Commands

Preliminary Basics

R is so popular for its vast array of packages available and a standard set of them is included. Packages are collections of R functions, data, and compiled codes in a well-defined format.

For the purpose of this book, we used a set of external packages.

The first step is to download and install a package and then, to use the package, invoke the *library* command to load it into the current session.

© The Author(s) 2018
A.A. Romano et al., *Climate Finance as an Instrument to Promote the Green Growth in Developing Countries*, SpringerBriefs in Climate Studies,
DOI 10.1007/978-3-319-60711-5

```
#to install package (once time)
install.packages("package")
#to load library of package (in each session)
library(package)
```

Sometime, developers can distribute R packages that are developing on GitHub. Thus, the *devtools* package provides *install_github()* that enables installing packages from GitHub (a development platform where to host code, manage projects, and build software jointly others developers).

```
#to install package from GitHub, first we need to install and
load devtools package
install.packages("devtools")
library(devtools)
#then, to install package from GitHub
devtools::install_github("repository_names/package")
```

Loading Data

The first step, and sometime more difficult, is to prepare and load data for successive operations.

The two more common formats of files are tab delimited text (*txt*) and comma separated values (*csv*), but equally common for statistical analysis are the spreadsheet file (*xls, xlsx*) and Stata dataset (*dta*).

Generally, the format of dataset is a matrix with the cases in row and variables in columns.

We assume that the file of dataset is in the current working directory.

```
#to show the current working directory
getwd()
#to set a different working directory must shall indicate the
full path, for example
setwd ("C:/Users/.../Data")
```

To load data from three different sources we must shall download packages and load libraries.

```
#to import data from a comma separated values
install.packages("readr") #Wickham et al. (2016)
library(readr)

#to import data from a Stata dataset
install.packages("haven") #Wickham and Miller. (2016)
library(haven)

#to import data from a spreadsheet
install.packages("readxl ") #Wickham (2016)
library(readxl)
```

Now we can import data in R environment.

```
#import data from a semi-column separated values (csv)
flow_disb <-
  read_delim(
    "flow_disb_comp.csv",
    ";",
    escape_double = FALSE,
    na = "NA",
    trim_ws = TRUE
  )
#NOTE: the option "escape_double" is to file escape quotes by
doubling them; "na" indicates the character vector of strings
to use for missing values; "trim_ws" indicates if whitespace
be trimmed from each field.

#to import data from a Stata dataset
new <- read_dta("new.dta")

#to import data from a spreadsheet
indic <- read_excel("indic.xlsx",
                    sheet = "indic")
```

It is recommendable to verify the imported data with *head()* and *tail()* function that return the first or last parts, respectively, of imported data.

```
head(new)
tail(new)
```

Sometime we need subset and order the dataset.

```
new_cat1<-subset(new[order(-new$Tot_rec, new$CC),], categ==1)
#NOTE: We created a subset of dataset "new" where the variable categ
is equal to 1 and it is ordered by the variable Tot_rec and CC
```

Exploring Data

We report the codes used to reproduce the figures in the text. The figures was mainly created with "plotly" package (Sievert et al. 2016). Thus, we need install and load the package.

```
install.packages("plotly")
#to install the latest version by github
devtools::install_github("ropensci/plotly") library(plotly)
```

Fig. 3.1 Committed and disbursed funding in terms of financing type

The Fig. 3.1 shows a pie chart with the sum of commitment and disbursement for every type of flow. Before plot the graph, we need to create a suitable dataset. We need a dataset where rows are the flows type to plot and the columns are the sum of commitment and disbursement. To do this we use a function to merge rows grouping by a category (*aggregate*) and rename the columns created (*setNames*). Both functions are included in *stats* package.

```
sum_comm_disb <-
#to insert in variable the new dataset
  setNames(
    aggregate(
      cbind(
#to merge needed columns
        flow_disb$commitment_amount_usd_constant,
```

(continued)

```
        flow_disb$disbursement_amount_usd_constant
        ),
        by = list(crs_flow_name = flow_disb$flow_type),
#the clause condition to group the rows
        FUN = sum,
#the function to apply to grouped rows
        na.rm = TRUE,
        na.action = NULL
#to remove the missing value and do not apply operations to
them
        ),
        c('crs_flow_name', 'comm', 'disb')
#the names to assign the new columns
    )
```

Now, we can proceed to plot the graph. We invite the reader to refer to official documentation of every package.

The figure will show two adjacent pie charts (delimited in the space by *domain* option). The first pie chart shows the amount of commitment grouped by flow type, while the second (added to first with *add_pie* function) shows the grouped disbursed funding. Both graphs are drawn with a hole and they show inside every part of the pie the real value and the percentage on total. The y-x axis are not drawn, such as ticks labels.

The *annotation* function is used to create the caption for every pie chart. To export the graph we use the *export* function of *plotly* package.

```
#to create the variable containing the plot
p1 <-
  plot_ly(
    sum_comm_disb,
    labels = ~ crs_flow_name,
    values = ~ comm,
    type = 'pie',
    textposition = 'inside',
    textinfo = 'percent+value',
    insidetextfont = list(color = '#FFFFFF'),
    marker = list(colors = colors,
                  line = list(color = '#FFFFFF', width = 1)),
    name = "Commitment",
    domain = list(x = c(0, 0.5), y = c(0, 0.9)),
```

(continued)

```
   showlegend = T,
   hole = 0.3,
   sort = F
) %>%
add_pie(
   sum_comm_disb,
   labels = ~ crs_flow_name,
   values = ~ disb,
   type = 'pie',
   textposition = 'inside',
   textinfo = 'percent+value',
   insidetextfont = list(color = '#FFFFFF'),
   marker = list(colors = colors,
                 line = list(color = '#FFFFFF', width = 1)),
   name = "Disbursement",
   domain = list(x = c(0.5, 1), y = c(0, 0.9)),
   showlegend = T,
   hole = 0.3,
   sort = F
) %>%
layout(
   title = 'Flow type in 2010 (% of Commitment & Disburse-
ment)',
   font = list(family = 'sans serif', size = 14),
   legend = list(
     x = 0.5,
     y = -0.10,
     xanchor = "center",
     orientation = 'h',
     font = list(family = 'sans serif', size = 12)
   ),
   showlegend = T,
   xaxis = list(
     showgrid = FALSE,
     zeroline = FALSE,
     showticklabels = FALSE
   ),
yaxis = list(
     showgrid = FALSE,
     zeroline = FALSE,
     showticklabels = FALSE
```

(continued)

```
  ),
  annotations = list(
    list(
      x = 0.15 ,
      y = -0.1,
      text = "Committed Funds",
      showarrow = F,
      xref = 'paper',
      yref = 'paper'
    ),
    list(
      x = 0.85 ,
      y = -0.1,
      text = "Disbursed Funds",
      showarrow = F,
      xref = 'paper',
      yref = 'paper'
    )
  )
)
#to display and check the output
p1

#to export the created pie chart in the working directory
export(p1, file = "flowtype.png")
```

Fig. 3.2 Committed and disbursed funding following the World Bank's Income classification of recipientcountries

The figure 3.2 shows a grouped bar chart to compare the amount of commitment and disbursement for every level of recipient's income. To have the total amount of both funding for every income's level we can create a suitable dataset as before done, or we can use an internal function of *plotly* package using another function belonging to *dplyr* package (Wickham and Francois 2016). Thus, we before install and load the package.

```
install.packages("dplyr ")
library(dplyr)
```

Now, we can proceed to plot the graph. The first argument is the dataset to use. The function *arrange*, belonging to *dplyr* package, allows to group the rows containing the individual disbursement and commitment (for each flow) by summing them for every income level and order them from high commitment and then high disbursement. To obtain a grouped by chart we use the function *add_bars*. The order of labels is fixed by two options: *type* and *categoryorder* (in layout section). The former indicates that the labels are passed by category of *x* values and the second indicates that the order is the same of dataset (which we have ordered by *tot_comm* and *tot_disb*).

```
p2 <-
  plot_ly(
    arrange(flow_disb%>%
      group_by(rec_income)%>%
      summarize(
        tot_comm=sum(commitment_amount_usd_constant),
        tot_disb=sum(disbursement_amount_usd_constant)
        ), desc(tot_comm, tot_disb)),
    x = ~ rec_income,
    y = ~ tot_comm,
    type = 'bar',
    name = 'Commitment'
  ) %>%
  add_bars(y = ~ tot_disb,
           name = 'Disbursement') %>%
  layout(
    title = "Amount of Commitment and Disbursement for Income
level of recipient",
    font = list(family = 'sans serif', size = 12),
    xaxis = list(
      title = "Income level of Recipient",
      x = 0 ,
      y = -0.5,
      tickfont = list(family = 'sans serif', size = 10),
      xref = 'paper',
      yref = 'paper',
      type = 'category',
      categoryorder = 'trace'
    ),
    yaxis = list(title = "Amount usd constant"),
    legend = list(
      x = 1,
      y = -0.3,
      xanchor = "right",
      orientation = 'h',
      font = list(family = 'sans serif', size = 10),
```

(continued)

```
      traceorder = "normal"
   )
 )

p2

export(p2, file = "Comm-Disb_Income.png")
```

Fig. 3.3 Incidence of Disbursement on pledges

The figure 3.3 shows a radar chart (a.k.a. spider plot) to represent the incidence of pledges of donors.

To draw it we use the function *radarchart* belonging to *fmsb* package (Nakazawa 2015). The dataset to be use must include maximum values as row 1, minimum values as row 2 and the data to show as row 3 for each variables, and the number of columns (variables) must be more than 2.

Then, to export the image as *ps* file, we use the function *CairoPS* belonging to *Cairo* package (Urbanek and Horner 2015).

```
#for radarchart function
install.packages("fmsb")
library(fmsb)

#for export image
install.packages("Cairo")
library(Cairo)
```

```
CairoPS(
  file = "pledge",
  bg = "white",
  width = 5,
  pointsize = 12
)
#radar chart
radarchart(
  indic,
  maxmin = T,
  axistype = 1,
  #custom the grid
```

(continued)

```
  cglcol = "grey80",
  cglty = 1,
  axislabcol = "grey10",
  cglwd = 0.5,
  #custom labels
  vlcex = 0.8,
  title = "Incidence of Disbursement"
)
dev.off()
```

Fig. 3.4 Geographic distribution of commitment and disbursement

The figure shows a grouped bar chart to compare the amount of commitment and disbursement for every recipient's geographic region. The code are similar to the code used to draw the Fig. 3.2.

```
p4 <-
  plot_ly(
    arrange(flow_disb%>%
                group_by(region)%>%
                summarize(
                  tot_comm=sum(commitment_amount_usd_constant),
                  tot_disb=sum(disbursement_amount_usd_constant)
                ), desc(tot_comm, tot_disb)),
    x = ~ region,
    y = ~ tot_comm,
    type = 'bar',
    name = 'Commitment'
  ) %>%
  add_bars(y = ~ tot_disb,
            name = 'Disbursement') %>%
  layout(
    title = "Amount of Commitment and Disbursement for Region
of recipient",
    font = list(family = 'sans serif', size = 12),
    xaxis = list(
      title = "Region of Recipient",
      x = 0 ,
      y = -0.5,
      tickfont = list(family = 'sans serif', size = 10),
      xref = 'paper',
```

(continued)

```
      yref = 'paper',
      type = 'category',
      categoryorder = 'trace'
   ),
   yaxis = list(title = "Amount usd constant"),
   legend = list(
      x = 1,
      y = -0.3,
      xanchor = "right",
      orientation = 'h',
      font = list(family = 'sans serif', size = 10),
      traceorder = "normal"
   )
 )
p4

export(p4, file = "Comm-Disb_Region.png")
```

Fig. 3.5 Flow of climate finance for the year 2010

The figure 3.5 shows the map of flow of funding, disbursed in 2010, in the World and in Europe.

First, we need to import data.

```
#Import Data
#dati_na contains the data related to EPI indicator (CC var-
iable) and category of country (cod indicates if it is a
donor, recipient or untreated country)
dati_na <-
  read.csv(
    "dataset_bioen.csv",
    header = T,
    sep = ";",
    dec = ".",
    na.strings = "NULL"
  )


#dati_flow contains the data related to each flow of funding
and includes the ratios of donors (and recipients) on total
amount of disbursed funding, the amount disbursed by each
donor and the latitude and longitude of donors and recipi-
ents.
```

(continued)

```
dati_flow <-
  read.csv(
    "flow_bio_en.csv",
    header = T,
    sep = ";",
    dec = ".",
    na.strings = "NULL"
  )
```

To draw the two map, colored on basis of the level of emissions (summarize by EPI) of each country we use the *rworldmap* package (South 2011), while to represent the flows of funding we use the function *gcIntermediate* belonging to *geosphere* package (Hijmans 2016), used to build another function *clean.Inter* (reported following). The colors of quantiles' distribution of EPI and category of countries (*cod*) come from the *classInt* package (Bivand 2015) (to obtain the interval of values within any quantile) and *RColorBrewer* package (Neuwirth 2014) to get the color palettes.

Then, to export the image as *PDF* file, we use the function *mapDevice* belonging to *rworldmap* package that creates a plot device suited for *rworldmap* plotting functions.

Thus, we start installing and loading the needed packages.

```
#to draw maps
install.packages("rworldmap")
library(rworldmap)

#to draw lines of flows
install.packages("geosphere")
library(geosphere)

#to get classes of quantiles
install.packages("classInt")
library(classInt)

#to get the color palettes
install.packages("RColorBrewer")
library(RColorBrewer)
```

Now, we can create the basis for draw the maps, starting with the building of needed functions.

```
#Function to draw lines of the flows
checkDateLine <- function(l) {
  n <- 0
  k <- length(l)
  k <- k - 1
  for (j in 1:k) {
    n[j] <- l[j + 1] - l[j]
  }
  n <- abs(n)
  m <- max(n, rm.na = TRUE)
  ifelse(m > 30, TRUE, FALSE)
}


clean.Inter <- function(p1, p2, n, addStartEnd) {
    inter <- gcIntermediate(p1, p2, n = n, addStartEnd =
addStartEnd)
    if (checkDateLine(inter[, 1])) {
    m1 <- midPoint(p1, p2)
    m1[, 1] <- (m1[, 1] + 180) %% 360 - 180
    a1 <- antipode(m1)
    l1 <- gcIntermediate(p1, a1, n = n, addStartEnd =
addStartEnd)
    l2 <- gcIntermediate(a1, p2, n = n, addStartEnd =
addStartEnd)
    l3 <- rbind(l1, l2)
    l3
  }
  else{
    inter
  }
}


# functions to add transparence to colors
addalpha <- function(colors, alpha = 1.0) {
  r <- col2rgb(colors, alpha = T)
  # Apply alpha
  r[4,] <- alpha * 255
  r <- r / 255.0
  return(rgb(r[1,], r[2,], r[3,], r[4,]))
}
```

The following text box include the code to join the imported data referenced by country codes (*iso_3* variable) to the internal map of *rworldmap* package.

```
#Create basis for map

set.seed(123)

sPDF <-
  joinCountryData2Map(
    dati_na ,
    joinCode = "ISO3" ,
    nameJoinColumn = "iso_3",
    nameCountryColumn = "country",
    suggestForFailedCodes = FALSE,
    mapResolution = "coarse",
    projection = NA,
    verbose = T
  )
#NOTE: the verbose option enables to check the successful
joins

#to remove the Antarctica
sPDF <- sPDF[-which(row.names(sPDF) == 'Antarctica'),]

#getting class intervals using a 'jenks' classification in
classInt package
classInt <-
  classIntervals(
    sPDF$CC,
    n = 4,
    style = "quantile",
    unique = T,
    dataPrecision = T,
    intervalClosure = c("left", "right")
  )
catMethod = classInt$brks

#getting a color scheme
#for quantiles
colourPalette <- brewer.pal(4, 'YlGn')
colourPalette <- addalpha(colourPalette, 0.5)
#for cod
```

(continued)

```
colourPalette1 <- colorRampPalette(c("grey60", "grey30"), space
= "Lab", bias = 3)(3)
colourPalette1 <- addalpha(colourPalette1, 0.3)
```

The following code shows the procedure to recreate the maps. The map are an overlapping of the two maps. In the former, the countries are hatched using *cod* variable. The second map is overlaid on the first and the countries are colored using the *CC* (EPI) variable.

```
#to open the graphic device
mapDevice(
    device = "pdf",
    file = "flowWorld.pdf",
    useDingbats = T,
    pagecentre = T,
    mai = c(0, 0, 0.2, 0),
    xaxs = "i",
    yaxs = "i"
)

#to set the parameters of first map
mapParams1 <-
    mapCountryData(
        sPDF,
        mapRegion = "world",
        nameColumnToPlot = 'cod',
        catMethod = "categorical",
        colourPalette = colourPalette1,
        addLegend = FALSE,
        borderCol = "black",
        mapTitle = "",
        aspect = 1,
        missingCountryCol = "grey90",
        lwd = 0.5,
        oceanCol = "aliceblue",
        nameColumnToHatch = "cod"
    )
#to set the parameters of second map
mapParams <-
    mapCountryData(
```

(continued)

```
   sPDF,
   mapRegion = "world",
   nameColumnToPlot = 'CC',
   catMethod = catMethod,
   colourPalette = colourPalette,
   addLegend = FALSE,
   borderCol = "black",
   mapTitle = "Flow of Funds in the World",
   aspect = 1,
   missingCountryCol = "grey90",
   add = T,
   lwd = 0.5,
   oceanCol = "aliceblue"
 )

#to add legend on map
do.call(
  addMapLegend
  ,
  c(
    mapParams,
    legendLabels = "all",
    legendWidth = 0.5,
    legendIntervals = "data" ,
    legendMar = 2
  )
)


#to draw lines of flows
#the width of lines are proportional to ratio of amount dis-
bursed by donor to a specific recipient country on total
amount disbursed by same donor to all recipients.

pal <- colorRampPalette(c("#ccffcc", "#006600"))
colors <- pal(100)

for (i in 1:length(dati_flow[, 1]))
{
  gC <-
    clean.Inter(dati_flow[i, c(4, 3)],
                dati_flow[i, c(7, 6)],
```

(continued)

```
                    n = 100,
                    addStartEnd = TRUE)

  colindex <-
    round((dati_flow[i,]$ratio_t_don) / 100 * length(colors))
  lines(gC, col = colors[colindex], lwd = (colindex / 50))

}


#to draw marker points of countries: red circle for donors
and green triangle point down for recipients
#the expansion of symbols are proportional to ratio of amount
disbursed by donor on total disbursed to all recipient coun-
tries for donor symbol and to ratio of amount received by
recipient on total funding for recipient symbol.

radius_don <- sqrt(dati_flow$ratio_gt_don / pi)
radius_rec <- sqrt(dati_flow$ratio_gt_rec / pi)

points(
  dati_flow$lon_don,
  dati_flow$lat_don,
  pch = 1,
  cex = radius_don ,
  col = "red"
)
points(
  dati_flow$lon_rec,
  dati_flow$lat_rec,
  pch = 6,
  cex = radius_rec ,
  col = "green"
)

legend(
  "bottomleft",
  box.col = "grey85",
  c("D", "T"),
  pch = c(1, 6),
  col = c("red", "green"),
  inset = c(.05, 0.2),
```

(continued)

```
   xjust = 0,
   cex = 0.9
)

dev.off()
```

Fig. 3.6 Flow of climate finance for the year 2010 with focus on European countries

The code included in the following box are similar to the code used to draw the previous map, but focus on Europe region.

```
mapDevice(
   device = "pdf",
   file = "flowEurope.pdf",
   useDingbats = T,
   pagecentre = T,
   mai = c(0, 0, 0.2, 0),
   xaxs = "i",
   yaxs = "i"
)

mapParams1 <-
   mapCountryData(
     sPDF,
     mapRegion = "eurasia",
     nameColumnToPlot = 'cod',
     catMethod = "categorical",
     colourPalette = colourPalette1,
     addLegend = FALSE,
     borderCol = "black",
     mapTitle = "",
     aspect = 1,
     missingCountryCol = "grey90",
     lwd = 0.5,
     oceanCol = "aliceblue",
     nameColumnToHatch = "cod"
   )

mapParams <-
   mapCountryData(
```

(continued)

```
      sPDF,
      mapRegion = "eurasia",
      nameColumnToPlot = 'CC',
      catMethod = catMethod,
      colourPalette = colourPalette,
      addLegend = FALSE,
      borderCol = "black",
      mapTitle = "Flow of Funds in Europe",
      aspect = 1,
      missingCountryCol = "grey90",
      add = T,
      lwd = 0.5,
      oceanCol = "aliceblue"
   )

do.call(
   addMapLegend,
   c(
      mapParams,
      legendLabels = "all",
      legendWidth = 0.5,
      legendIntervals = "data",
      legendMar = 2
   )
)

# flow lines #

pal <- colorRampPalette(c("#ccffcc", "#006600"))
colors <- pal(100)

for (i in 1:length(dati_flow[, 1]))
{
   gC <-
      clean.Inter(dati_flow[i, c(4, 3)],
                  dati_flow[i, c(7, 6)],
                  n = 100,
                  addStartEnd = TRUE)
```

(continued)

```r
  colindex <-
    round((dati_flow[i,]$ratio_t_don) / 100 * length(colors))
  lines(gC, col = colors[colindex], lwd = (colindex / 50))


}


radius_don <- sqrt(dati_flow$ratio_gt_don / pi)
radius_rec <- sqrt(dati_flow$ratio_gt_rec / pi)

# marker points #

points(
  dati_flow$lon_don,
  dati_flow$lat_don,
  pch = 1,
  cex = radius_don ,
  col = "red"
)
points(
  dati_flow$lon_rec,
  dati_flow$lat_rec,
  pch = 6,
  cex = radius_rec ,
  col = "green"
)

legend(
  "bottomleft",
  box.col = "grey85",
  c("D", "T"),
  pch = c(1, 6),
  col = c("red", "green"),
  inset = c(.05, 0.2),
  xjust = 0,
  cex = 0.9
)

dev.off()
```

Fig. 3.7 Number of Treated and Untreated Countries on the basis of quartiles of GHG emission distribution

The figure 3.7 shows a stacked horizontal bar chart to compare the number of treated and untreated countries for every quantile of EPI's distribution using the function *count*. The code are similar to the code used to draw the Figs. 3.2, 3.4.

```
p7 <- new %>% count(categ, d_t_u) %>%
  plot_ly(
    x = ~ n,
    y = ~ categ,
    type = "bar",
    orientation = 'h',
    textposition = 'inside',
    text = ~ n ,
    insidetextfont = list(color = '#FFFFFF'),
    color = ~ d_t_u
  ) %>%
  layout(
    barmode = 'stack',
    title = "Total of Treated and Untreated countries",
    font = list(family = 'sans serif', size = 12),
    yaxis = list(
      title = "Quantiles of GHG distribution",
      tickmode = "array",
      tickvals = 1:5,
      ticktext = c("25th", "50th", "75th", "90th", "Over 90th")
    ),
  xaxis = list(title = "Number of Countries"),
  legend = list(
  x = 1,
  y = -0.10,
  xanchor = "right",
  orientation = 'h',
  font = list(family = 'sans serif', size = 10),
  traceorder = "normal"
  )
 )
p7
export(p7, file = "treatUnt.png")
```

Fig. 3.8 Climate funds received by developing countries

The figure 3.8 shows a horizontal bar chart to compare the amount of disbursed funding for every quantile of EPI's distribution. The code are similar to the code used to draw the Fig. 3.7.

```
p8 <- plot_ly(
  new,
  x = ~ Tot_rec,
  y = ~ categ,
  type = "bar",
  orientation = 'h',
  textposition = 'inside',
  insidetextfont = list(color = '#FFFFFF')
) %>%
  layout(
    barmode = 'stack',
    title = "Total of Recipients' received funding",
    font = list(family = 'sans serif', size = 12),
    yaxis = list(
      title = "Quantiles of GHG distribution",
      tickmode = "array",
      tickvals = 1:5,
      ticktext = c("25th", "50th", "75th", "90th", "Over 90th")
    ),
    xaxis = list(title = "$ Disbursed"),
    legend = list(
    x = 1,
    y = -0.10,
    xanchor = "right",
    orientation = 'h',
    font = list(family = 'sans serif', size = 10),
    traceorder = "normal"
    )
  )
p8
export(p81, file = "treat_disb.png")
```

Fig. 3.9 Disbursement and Environmental Performance of recipients

The figure 3.9 shows two-stacked horizontal bar chart with a line plot to compare the amount of funding with level of GHG emissions (EPI) and with share of renewable/fossil source, for the countries included in 25th quantile of EPI's distribution. To build the figure we use the function *subplot* belonging to *plotly* package whit the option "*shareY*" to align every y-axis and option "*legendgroup*" to group the entry legend for every plot. For the code of bar chart, see Fig. 3.7.

The order of labels on y-axis is indicated by the option "*categoryorder = "array"*" where array include the names of countries ordered by amount of funding, first, and level of GHG emission, after.

```
p9.1 <-
  plot_ly(
    data = new_cat1,
    x = ~ tot_rec_en,
    y = ~ country,
    type = 'bar',
    orientation = 'h',
    name = 'Energy',
    marker = list(color = 'green', line = list(color = 'green',
width = 1)),
  legendgroup = 'group1'
  ) %>%
  add_trace(
   x = ~ tot_rec_bio,
   name = 'Biosphere',
   marker = list(
    color = 'deepskyblue',
    line = list(color = 'deepskyblue', width = 1)
   ),
   legendgroup = 'group1'
  ) %>%
  layout(
   barmode = 'stack',
   yaxis = list(
    type = "category",
    categoryorder = "array",
    categoryarray = new_cat1$country[with(new_cat1, order
(Tot_rec, -CC))],
    showgrid = TRUE,
    showline = FALSE,
    showticklabels = TRUE,
    domain = c(0, 0.9)
   ),
   xaxis = list(
    zeroline = FALSE,
    showline = FALSE,
```

(continued)

```
      showticklabels = TRUE,
      showgrid = TRUE
    )


  )
p9.2 <-
  plot_ly(
    data = new_cat1,
    x =    ~ CC,
    y =    ~ country,
    type = 'scatter',
    name = 'Ghg emissions',
    mode = 'lines+markers',
    line = list(color = 'rgb(128, 0, 128)'),
    legendgroup = 'group2'
  ) %>%
  layout(
    yaxis = list(
      type = "category",
      categoryorder = "array",
      categoryarray = new_cat1$country[with(new_cat1, order
(Tot_rec,-CC))],
      showgrid = TRUE,
      showline = TRUE,
      showticklabels = FALSE,
      linecolor = 'rgba(102, 102, 102, 0.8)',
      linewidth = 2,
      domain = c(0, 0.9)
    ),
    xaxis = list(
      zeroline = FALSE,
      showline = FALSE,
      showticklabels = TRUE,
      showgrid = TRUE
    )
  )
p9.3 <-
  plot_ly(
    data = new_cat1,
    x = ~ sh_foss,
    y = ~ country,
```

(continued)

```
     type = 'bar',
     orientation = 'h',
     name = 'Share of Fossil',
     marker = list(color = 'rgba(204,204,204,1)', line = list
(color = 'grey', width = 0.5)),
     legendgroup = 'group3'
   ) %>%
  add_trace(
     x = ~ sh_ren,
     type = 'bar',
     orientation = 'h',
     name = 'Share of Renewable',
     marker = list(color = 'rgba(50, 171, 96, 0.7)', line = list
(color = 'green', width = 0.5)),
   legendgroup = 'group3'
 ) %>%
 layout(
  barmode = 'stack',
  yaxis = list(
   type = "category",
   categoryorder = "array",
   categoryarray   =   new_cat1$country[with(new_cat1,   order
(Tot_rec,-CC))],
   showgrid = TRUE,
   showline = FALSE,
   showticklabels = TRUE,
   domain = c(0, 0.9)
  ),
  xaxis = list(
   zeroline = FALSE,
   showline = FALSE,
   showticklabels = TRUE,
   showgrid = TRUE
  )
 )

p9.1_2_3 <- subplot(p9.1, p9.2, p9.3, shareY = TRUE, margin =
0) %>%
  layout(
    autosize = T,
    margin(t = 0),
```

(continued)

```
title = "Total funds and Environmental Performance",
font = list(family = 'sans serif', size = 14),
legend = list(
  xref = 'paper',
  yref = 'paper',
  x = 1,
  y = -0.1,
  xanchor = "right",
  orientation = 'h',
  font = list(family = 'sans serif', size = 10)
),
yaxis = list(
  title = '',
  tickangle = 45,
  tickfont = list(family = 'sans serif', size = 8)
),
xaxis = list(
  title = '',
  tickangle = 45,
  tickfont = list(family = 'sans serif', size = 8)
),
xaxis2 = list(
  title = '',
  tickangle = 45,
  tickfont = list(family = 'sans serif', size = 8)
),
xaxis3 = list(
  title = '',
  tickangle = 45,
  tickfont = list(family = 'sans serif', size = 8)
)
) %>%
add_annotations(
  xref = 'paper',
  yref = 'paper',
  x = 0,
  y = -0.15,
  text = paste('25% quantile of GHG distribution'),
  font = list(
    family = 'sans serif',
    size = 10,
```

(continued)

```
     color = 'rgb(150,150,150)'
  ),
   showarrow = FALSE
 )

p9.1_2_3

export(p9.1_2_3, file = "tot_funds_and_envir_perfor.png")
```

Fig. 3.10 Disbursement and Development of recipients

The figure 3.10 shows two-stacked horizontal bar chart to compare the amount of funding with level of GDP and the Electricity consumption, for the countries included in 25th quantile of EPI's distribution. The code is similar to the code of Fig. 3.9.

```
p10 <-
  plot_ly(
    data = new_cat1,
    y = ~ country,
    x = ~ lnelcons,
    type = 'bar',
    orientation = 'h',
    name = 'Electricity Consumption (log scale)',
    marker = list(color = 'rgba(222, 237, 255, 0.7)', line =
list(color = 'green', width = 0.5)),
    legendgroup = 'group3'
  ) %>%
  add_trace(
    x = ~ lgdp,
    type = 'bar',
    orientation = 'h',
    name = 'GDP (log scale)',
    marker = list(color = 'rgba(150,150,150,1)', line = list
(color = 'grey', width = 0.5)),
    legendgroup = 'group3'
  ) %>%
  layout(
    barmode = 'group',
    yaxis = list(
      type = "category",
```

(continued)

```
      categoryorder = "array",
      categoryarray = new_cat1$country[with(new_cat1, order
(Tot_rec,-CC))],
      showgrid = TRUE,
      showline = FALSE,
      showticklabels = TRUE,
      domain = c(0, 0.9)
    ),
    xaxis = list(
      zeroline = FALSE,
      showline = FALSE,
      showticklabels = TRUE,
      showgrid = TRUE
    )
  )
p10_9.1_4 <- subplot(p9.1, p10, shareY = TRUE) %>%
  layout(
    title = "Total funds and Development",
    font = list(family = 'sans serif', size = 12),
    legend = list(
      x = 1,
      y = -0.10,
      xanchor = "right",
      orientation = 'h',
      font = list(family = 'sans serif', size = 10),
      traceorder = "grouped+reversed"
    ),
    yaxis = list(
      title = '',
      tickangle = 45,
      tickfont = list(family = 'sans serif', size = 8)
    ),
    xaxis = list(
      title = '',
      tickangle = 45,
      tickfont = list(family = 'sans serif', size = 8)
    ),
    xaxis2 = list(
      title = '',
      tickangle = 45,
      tickfont = list(family = 'sans serif', size = 8)
```

(continued)

```
      )
    ) %>%
    add_annotations(
      xref = 'paper',
      yref = 'paper',
      x = 0,
      y = -0.3,
      text = paste('25% quantile of GHG distribution'),
      font = list(
        family = 'sans serif',
        size = 8,
        color = 'rgb(150,150,150)'
      ),
      showarrow = FALSE
    )

p10_9.1_4

export( p10_9.1_4, file = "tot_funds_and_dev.png")
```

Building the Indicator

To obtain the Composite Indicators of GHG emission with the weighting method based on factor analysis, we used the function *ci_factor* included in *Compind* package (Vidoli et al. 2015). The raw data of GHG are in the dataset *dataCC*.

```
CI<-ci_factor(dataCC,method="CH", dim=2)
# method = "CH" can be choose the number of the component to
take into account while dim=2 indicates the number of chosen
component
```

Now, we can import the Composite indicator estimated values in dataset.

```
data$CI<-CI$ci_factor_est
```

To scale values between [0; 1] corresponding to [min; max], it's straight-forward to create a small function to do this using basic arithmetic:

```
range01 <- function(x){(ma(x)-x)/(max(x)-min(x))}
```

and then we can import in dataset the scaled values to compare the results.

```
dataCC$CI_norm<-range01(CI$ci_factor_est)
```

Lastly, we can export the new dataset in working directory using the function *write.xlsx* included in *xlsx* package (Dragulescu 2014).

```
install.packages("xlsx")
library(xlsx)

#to export the dataset
write.xlsx(dataCC, "dataCC_new.xlsx")
```

STATA Commands

In this book, we estimated a quantile regression using STATA software, because it wasn't implemented in R software a package with the Parente-Santo Silva procedure. Thus, we estimated the parameters of quantile regression using the *qreg2* module (Parente and Santos Silva 2016) for every decile.

```
//to install the qreg2 module
ssc install qreg2
```

```
# Execute quantile regression for every quantile and store
results
# With the option cluster(GroupI) we specify that the stan-
dard errors are computed allowing for intra-cluster correla-
tion as in Parente and Santos Silva (2016) to account the
social-economic structure of countries (GroupI indicates the
income group of WorldBank).
```

(continued)

```
qreg2 CC Tot_rec pop_fem ei oil_sup sh_foss sh_n lgdp
lnelcons acc_el , quantile(.10) c(GroupI)
est store CC_10th_tot
qreg2 CC tot_rec_en pop_fem ei oil_sup sh_foss sh_n lgdp
lnelcons acc_el , quantile(.10) c(GroupI)
est store CC_10th_en
qreg2 CC tot_rec_bio pop_fem ei oil_sup sh_foss sh_n lgdp
lnelcons acc_el , quantile(.10) c(GroupI)
est store CC_10th_bio

// Execute linear regression specifying that the standard
error reported allow for intragroup correlation and then
store results
regress CC Tot_rec pop_fem ei oil_sup sh_foss sh_n lgdp
lnelcons acc_el, vce(cluster GroupInc)
est store CC_OLS_tot
```

We can export the stored results in a rtf file, to manage it and import them in R software.

```
// export results (this sis only for the total disbursed)
esttab  CC_OLS_tot   CC_10th_tot   CC_20th_tot   CC_30th_tot
CC_40th_tot CC_50th_tot CC_60th_tot CC_70th_tot CC_80th_tot
CC_90th_tot  CC_95th_tot  using  C:\...\results.rtf,  append
title(This is a quantile regression table Total) nonumbers
mtitles("OLS" "10th" "20th" "30th" "40th" "50th" "60th"
"70th" "80th" "90th" "95th")cells(b(star label(Coef.) fmt
(a3)) se(par fmt(a3))) label varlabels(_cons Constant) legend
stats(r2 N pss_p, labels(R-squared "N. of cases" "Parente-
Santos Silva test p-value")) starlevels(* 0.1 ** 0.05 ***
0.01)

// export coefficients and confidence intervals to be used in R
software
estout  CC_OLS_tot   CC_10th_tot   CC_20th_tot   CC_30th_tot
CC_40th_tot CC_50th_tot CC_60th_tot CC_70th_tot CC_80th_tot
CC_90th_tot CC_95th_tot using C:\...\results_b.xls, cells(b)
estout  CC_OLS_tot   CC_10th_tot   CC_20th_tot   CC_30th_tot
CC_40th_tot CC_50th_tot CC_60th_tot CC_70th_tot CC_80th_tot
CC_90th_tot CC_95th_tot using C:\...\results_ci.xls, cells
(ci_1 ci_u)
```

R: Create graph of Quantiles Plot (Fig. 5.3 (a) – (j) Quantile coefficient plots)

To create quantile coefficients plot we can use the STATA command, but it is not implemented to obtain confidence interval for the Parente-Silva procedure. To remedy the lack of a specific procedure to obtain the confidence interval with the standard errors proposed by Parente-Silva, we can import the values of coefficients and the related lower and upper confidence intervals in R, after wwe exported them by STATA and create the plots using a simple loop in R.

```
# Import saved coefficients from Stata
data_coef <-
  data.frame(
    en_Tot_rec = c(..., ..., ... ),
    en_min_Tot_rec = c(..., ..., ...),
    en_max_Tot_rec = c(..., ..., ...),
    en_pop_fem = c( ... ) ...
          )
```

To create the loop we need to create a list of variables' names to pass in loop.

```
# create a list of variables' names
var_q <-
  c(
    "Tot_rec",
    "pop_fem",
    "ei",
    "oil_sup",
    "sh_foss",
    "sh_nonhydro",
    "lgdp",
    "lnelcons",
    "acc_el",
    "Constant"
  )
```

Now, we can construct the loop.

```
# loop for create and save plots

for (i in seq_along(var_q)) {
  CairoPS(
    file = var_q[i],
    bg = "white",
    width = 5,
    pointsize = 12
  )

  plot(
    q,
    data_coef[, paste("en_", var_q[i], sep = "")],
    type = 'n',
    ylim = c(min(data_coef[, paste("en_min_", var_q[i], sep =
"")], data_coef[, paste("bio_min_", var_q[i], sep =
                                "")]),
        max(data_coef[, paste("en_max_", var_q[i], sep =
                    "")], data_coef[, paste("bio_max_", var_q
[i], sep = "")])),
    xaxt = 'n',
    xlab = "quantile",
    ylab = var_q[i]
  )

  polygon(
    c(q, rev(q)),
    c(data_coef[, paste("en_max_", var_q[i], sep = "")], rev
(data_coef[, paste("en_min_", var_q[i], sep ="")])),
    col = SetAlpha("green4", .5),
    border = NA,
    density = 25,
    angle = 45,
    lty = 2
  )
  polygon(
    c(q, rev(q)),
    c(data_coef[, paste("bio_max_", var_q[i], sep = "")], rev
(data_coef[, paste("bio_min_", var_q[i], sep =
                                "")])),
    col = SetAlpha("blue", .5),
    border = NA,
    density = 25,
```

(continued)

```
   angle = -45,
   lty = 3
)

points(
   q,
   data_coef[, paste("en_", var_q[i], sep = "")],
   type = 'l',
   col = 'green4',
   lwd = 2,
   ylim = c(min(data_coef[, paste("en_min_", var_q[i], sep =
"")]), max(data_coef[, paste("en_max_", var_q[i], sep =
                              "")]))
)
points(
  q,
  data_coef[, paste("bio_", var_q[i], sep = "")],
  type = 'l',
  col = 'blue',
  lwd = 2,
   ylim = c(min(data_coef[, paste("bio_min_", var_q[i], sep =
"")]), max(data_coef[, paste("bio_max_", var_q[i], sep =
                              "")]))
)

   abline(
     h = data_coef[, paste("en_ols_", var_q[i], sep = "")],
     v = 0,
     col = 'green2',
     lty = 3,
     lwd = 1,
     ylim = c(min(data_coef[, paste("en_ols_min_", var_q[i],
sep = "")]), max(data_coef[, paste("en_ols_max_", var_q[i],
sep = "")]))
   )

   abline(
     h = data_coef[, paste("bio_ols_", var_q[i], sep = "")],
     v = 0,
     col = 'deepskyblue',
     lty = 3,
     lwd = 1,
```

(continued)

```
    ylim = c(min(data_coef[, paste("bio_ols_min_", var_q[i],
sep =
                    "")]), max(data_coef[, paste("bio_ols_max_",
var_q[i], sep = "")]))
  )


  abline(v = 0,
         h = 0,
         col = "grey30",
         lty = 6)

  axis(1, at = q)
  legend(
    "bottomright",
    c("Energy", "Bio", "OLS_en", "OLS_bio"),
    col = c("green4", "blue", "green2", "deepskyblue"),
    pch = 20
  )
  dev.off()
}
```

References

Bivand R (2015) classInt: choose univariate class intervals. R package version 0.1–23

Dragulescu AA (2014) xlsx: read, write, format excel 2007 and excel 97/2000/XP/2003 files. R package version 0.5.7

Hijmans RJ (2016) geosphere: spherical trigonometry. R package version 1.5–5

Nakazawa M (2015) fmsb: functions for medical statistics book with some demographic data. R package version 0.5.2

Neuwirth E (2014) RColorBrewer: colorBrewer palettes. R package version 1.1–2

Parente PM, Santos Silva J (2016) Quantile regression with clustered data. J Econ Methods 5 (1):1–15

R Core Team (2016) R: a language and environment for statistical computing. R Foundation for Statistical Computing, Vienna

Sievert C, Parmer C, Hocking T, Chamberlain S, Ram K., Corvellec M. Despouy P (2016) plotly: create interactive web graphics via 'plotly.js'

South A (2011) rworldmap: a new R package for mapping global data. The R J 3/1: 35–43

Urbanek S, Horner J (2015) Cairo: r graphics device using cairo graphics library for creating high-quality bitmap (PNG, JPEG, TIFF), vector (PDF, SVG, PostScript) and display (X11 and Win32) output. R package version 1.5–9

Vidoli F, Fusco E, Mazziotta C (2015) Non-compensability in composite indicators: a robust directional frontier method. Soc Indic Res 122(3):635–652

Wickham H (2016) readxl: Read excel files. R package version 0.1.1

Wickham H, Francois R (2016) dplyr: a grammar of data manipulation. R package version 0.5.0

Wickham H, Miller E (2016) haven: import and export 'SPSS', 'Stata' and 'SAS' files. R package version 1.0.0

Wickham H, Hester J, Francois R (2016) readr: read Tabular Data. R package version 1.0.0

Appendix B

List of Countries included in our study. For each countries is indicated the Iso3 code, its position on Climate Finance (if Donor, Treated or Untreated) and the Environmental Pollution Index (EPI).

Country	Iso 3	Donor/Treated/Untreated	EPI
Afghanistan	AFG	Treated	0.99254
Albania	ALB	Untreated	0.998725
Algeria	DZA	Treated	0.975688
Angola	AGO	Treated	0.962244
Antigua and Barbuda	ATG	Untreated	0.999436
Argentina	ARG	Treated	0.943719
Armenia	ARM	Treated	0.99853
Australia	AUS	Donor	0.889048
Austria	AUT	Donor	0.992426
Azerbaijan	AZE	Treated	0.987959
Bahamas, The	BHS	Untreated	0.999757
Bahrain	BHR	Untreated	0.997795
Bangladesh	BGD	Treated	0.966298
Barbados	BRB	Untreated	0.999367
Belarus	BLR	Treated	0.987373
Belgium	BEL	Donor	0.989001
Belize	BLZ	Untreated	0.997443
Benin	BEN	Treated	0.997723
Bhutan	BTN	Untreated	0.999816
Bolivia	BOL	Treated	0.989949
Bosnia and Herzegovina	BIH	Treated	0.997229
Botswana	BWA	Treated	0.99674
Brazil	BRA	Treated	0.784867
Brunei	BRN	Untreated	0.996453
Bulgaria	BGR	Untreated	0.993349
Burkina Faso	BFA	Untreated	0.993025
Burundi	BDI	Untreated	0.998966
Cambodia	KHM	Treated	0.993006

(continued)

Country	Iso 3	Donor/Treated/Untreated	EPI
Cameroon	CMR	Treated	0.967426
Canada	CAN	Donor	0.914518
Cape Verde	CPV	Treated	0.999902
Central African Republic	CAF	Untreated	0.981787
Chad	TCD	Untreated	0.990779
Chile	CHL	Treated	0.989776
China	CHN	Treated	0
Colombia	COL	Treated	0.970694
Comoros	COM	Untreated	0.999924
Costa Rica	CRI	Treated	0.997526
Cote d'Ivoire	CIV	Untreated	0.99284
Croatia	HRV	Untreated	0.996672
Cuba	CUB	Treated	0.992686
Cyprus	CYP	Untreated	0.999407
Czech Republic	CZE	Donor	0.990146
Denmark	DNK	Donor	0.993328
Djibouti	DJI	Untreated	0.999675
Dominica	DMA	Untreated	0.999965
Dominican Republic	DOM	Treated	0.995425
Ecuador	ECU	Treated	0.992697
Egypt	EGY	Treated	0.969811
El Salvador	SLV	Treated	0.99757
Equatorial Guinea	GNQ	Untreated	0.995627
Eritrea	ERI	Treated	0.997923
Estonia	EST	Untreated	0.998036
Ethiopia	ETH	Treated	0.961461
Fiji	FJI	Untreated	0.999565
Finland	FIN	Donor	0.992837
France	FRA	Donor	0.939076
Gabon	GAB	Untreated	0.999379
Gambia, The	GMB	Untreated	0.998068
Georgia	GEO	Untreated	0.99703
Germany	DEU	Donor	0.92129
Ghana	GHA	Treated	0.994691
Greece	GRC	Donor	0.991673
Grenada	GRD	Untreated	0.999523
Guatemala	GTM	Treated	0.994989
Guinea	GIN	Untreated	0.995367
Guinea-Bissau	GNB	Untreated	0.999355
Guyana	GUY	Untreated	0.999269
Haiti	HTI	Treated	0.997913
Honduras	HND	Treated	0.995995
Hungary	HUN	Untreated	0.991939

(continued)

Country	Iso 3	Donor/Treated/Untreated	EPI
Iceland	ISL	Donor	0.999555
India	IND	Treated	0.560014
Indonesia	IDN	Treated	0.882877
Iran	IRN	Treated	0.943113
Iraq	IRQ	Untreated	0.962057
Ireland	IRL	Donor	0.990769
Israel	ISR	Untreated	0.982969
Italy	ITA	Donor	0.953287
Jamaica	JAM	Untreated	0.998735
Japan	JPN	Donor	0.871739
Jordan	JOR	Treated	0.996996
Kazakhstan	KAZ	Treated	0.975621
Kenya	KEN	Treated	0.982847
Kiribati	KIR	Untreated	0.999993
Korea, Dem. Rep. (North)	PRK	Untreated	0.988797
Korea, Rep. (South)	KOR	Donor	0.929427
Kuwait	KWT	Untreated	0.963969
Kyrgyzstan	KGZ	Treated	0.997889
Laos	LAO	Treated	0.997015
Latvia	LVA	Untreated	0.997346
Lebanon	LBN	Treated	0.997793
Lesotho	LSO	Treated	0.99938
Liberia	LBR	Untreated	0.999569
Libya	LBY	Untreated	0.975507
Lithuania	LTU	Untreated	0.996133
Luxembourg	LUX	Donor	0.999293
Macedonia, FYR	MKD	Untreated	0.997388
Madagascar	MDG	Treated	0.991808
Malawi	MWI	Treated	0.997034
Malaysia	MYS	Treated	0.964568
Maldives	MDV	Treated	0.999943
Mali	MLI	Treated	0.990379
Malta	MLT	Untreated	0.999491
Mauritania	MRT	Untreated	0.997187
Mauritius	MUS	Untreated	0.999213
Mexico	MEX	Treated	0.881368
Moldova	MDA	Untreated	0.998261
Mongolia	MNG	Treated	0.992912
Montenegro	MNE	Treated	0.99908
Morocco	MAR	Treated	0.991569
Mozambique	MOZ	Treated	0.991746
Namibia	NAM	Treated	0.995872
Nepal	NPL	Treated	0.990549

(continued)

Country	Iso 3	Donor/Treated/Untreated	EPI
Netherlands	NLD	Donor	0.982099
New Zealand	NZL	Donor	0.983135
Nicaragua	NIC	Treated	0.996674
Niger	NER	Treated	0.991823
Nigeria	NGA	Treated	0.93274
Norway	NOR	Donor	0.994249
Oman	OMN	Untreated	0.986424
Pakistan	PAK	Tretated	0.938314
Panama	PAN	Treated	0.99757
Papua New Guinea	PNG	Treated	0.996202
Paraguay	PRY	Treated	0.98938
Peru	PER	Treated	0.985482
Philippines	PHL	Treated	0.971461
Poland	POL	Donor	0.96607
Portugal	PRT	Donor	0.992968
Qatar	QAT	Untreated	0.99559
Russian Federation	RUS	Untreated	0.71649
Rwanda	RWA	Treated	0.998085
Saint Kitts and Nevis	KNA	Untreated	0.999955
Saint Lucia	LCA	Untreated	0.999777
Saint Vincent and Grenadines	VCT	Untreated	0.999973
Samoa	WSM	Untreated	0.999911
Sao Tome and Principe	STP	Treated	0.999973
Saudi Arabia	SAU	Untreated	0.96773
Senegal	SEN	Treated	0.994075
Serbia	SRB	Treated	0.994221
Seychelles	SYC	Untreated	0.999947
Sierra Leone	SLE	Untreated	0.998303
Singapore	SGP	Untreated	0.99369
Slovakia	SVK	Donor	0.996441
Slovenia	SVN	Donor	0.998071
Solomon Islands	SLB	Untreated	0.999893
South Africa	ZAF	Treated	0.951351
Spain	ESP	Donor	0.96571
Sri Lanka	LKA	Treated	0.992422
Sudan	SDN	Untreated	0.953046
Suriname	SUR	Untreated	0.999474
Swaziland	SWZ	Untreated	0.999413
Sweden	SWE	Donor	0.993315
Switzerland	CHE	Donor	0.995168
Syria	SYR	Treated	0.98728
Tajikistan	TJK	Treated	0.997561
Tanzania	TZA	Treated	0.978226

(continued)

Country	Iso 3	Donor/Treated/Untreated	EPI
Thailand	THA	Treated	0.954946
Togo	TGO	Untreated	0.998476
Tonga	TON	Treated	0.999926
Trinidad and Tobago	TTO	Untreated	0.998105
Tunisia	TUN	Treated	0.995896
Turkey	TUR	Treated	0.960223
Turkmenistan	TKM	Untreated	0.985629
Uganda	UGA	Treated	0.989767
Ukraine	UKR	Treated	0.957511
United Arab Emirates	ARE	Untreated	0.982603
United Kingdom	GBR	Donor	0.946788
United States	USA	Donor	0
Uruguay	URY	Treated	0.990651
Uzbekistan	UZB	Untreated	0.959177
Vanuatu	VUT	Treated	0.999811
Venezuela	VEN	Treated	0.964524
Vietnam	VNM	Treated	0.96563
Yemen	YEM	Treated	0.995981
Zambia	ZMB	Treated	0.982392
Zimbabwe	ZWE	Untreated	0.993995

Printed in the United States
By Bookmasters